Recent Advancement in Microwave Photonics

Recent Advancement in Microwave Photonics

Guest Editors

Lei Zhang
Lu Wang
Tengfei Hao

MDPI

Basel • Beijing • Wuhan • Barcelona • Belgrade • Novi Sad • Cluj • Manchester

Guest Editors
Lei Zhang
State Key Laboratory of Information Photonics and Optical Communications
Beijing University of Posts and Telecommunications
Beijing
China

Lu Wang
Institute of Semiconductors
Chinese Academy of Sciences
Beijing
China

Tengfei Hao
Institute of Semiconductors
Chinese Academy of Sciences
Beijing
China

Editorial Office
MDPI AG
Grosspeteranlage 5
4052 Basel, Switzerland

This is a reprint of the Special Issue, published open access by the journal *Photonics* (ISSN 2304-6732), freely accessible at: https://www.mdpi.com/journal/photonics/special_issues/42QUDNSISD.

For citation purposes, cite each article independently as indicated on the article page online and as indicated below:

Lastname, A.A.; Lastname, B.B. Article Title. *Journal Name* **Year**, *Volume Number*, Page Range.

ISBN 978-3-7258-8115-4 (Hbk)
ISBN 978-3-7258-8116-1 (PDF)
https://doi.org/10.3390/books978-3-7258-8116-1

Cover image courtesy of Lei Zhang

Contents

Preface

The field of microwave photonics has witnessed remarkable growth in the last decade, evolving from a niche research area into a cornerstone technology for modern high-speed information systems. This Reprint, entitled "Recent Advancement in Microwave Photonics", aims to provide a comprehensive and scholarly platform showcasing the latest theoretical innovations and experimental breakthroughs in this interdisciplinary domain. By merging the vast bandwidth of photonics with the sophisticated signal processing capabilities of microwave engineering, the research presented here addresses the urgent need for systems that can operate with higher frequencies, lower latency, and reduced electromagnetic interference.

The scope of this work is multifaceted, ranging from the optimization of fundamental silicon-based and silicon nitride-based integrated components to the implementation of high-sensitivity phase noise analyzers and advanced radar architectures. These contributions represent the collective efforts of leading researchers to push the boundaries of what is possible in optical–electrical signal generation, modulation, and detection. This Reprint is addressed to a broad audience, including academic researchers, graduate students, and industry engineers who are seeking to understand the current state of the art and the future trajectory of photonic integrated circuits (PICs).

Looking forward, microwave photonics is poised to play a transformative role in the next generation of communication infrastructures. As we move toward the 6G era and expanded satellite-based communication networks, the demand for ultra-wideband signal processing and seamless space–air–ground integration will become paramount. Microwave photonics is uniquely positioned to meet these challenges by providing high-fidelity signal distribution and processing in both terrestrial and orbital environments.

Furthermore, the future of this field lies in the exploration of heterogeneous integration and the adoption of novel material platforms. The integration of ferroelectric materials, such as Barium Titanate (BTO), with silicon or silicon nitride substrates offers an exciting path toward achieving ultra-efficient, high-speed modulation with low power consumption. By harnessing the Pockels effect in such thin-film materials, next-generation microwave photonic systems will achieve unprecedented performance levels. We hope that the insights shared in this Reprint will inspire further innovation and serve as a catalyst for the development of future high-performance microwave photonic systems.

Lei Zhang, Lu Wang, and Tengfei Hao
Guest Editors

photonics

MDPI

Article

Low-Loss Silicon Nitride Bent Waveguides at O-Band with Modified Hermite Curves

Donghao Li [1], Enge Zhang [1,†], Yi Xu [1], Lei Zhang [1], Yu Zhang [2] and Xu Yang [2,*]

[1] School of Integrated Circuits, Beijing University of Posts and Telecommunications, Beijing 100876, China; ldhao@bupt.edu.cn (D.L.); zhangenge@bupt.edu.cn (E.Z.); xuyi@bupt.edu.cn (Y.X.); zhang-lei@bupt.edu.cn (L.Z.)

[2] The 54th Research Institute of China Electronics Technology Group Corporation, Shijiazhuang 050081, China; zyzhangyu1996@163.com

* Correspondence: yangxumail@foxmail.com

† Current address: School of Integrated Circuits, Ludong University, Yantai 264025, China.

Abstract

The silicon nitride (SiN) platform offers a low-loss solution for photonic integrated circuits (PICs). The bent waveguide represents one of the primary sources of propagation loss in such integrated systems. In this work, we present a comprehensive review of recent advances in SiN bent waveguide technologies. We propose a low-loss SiN bent waveguide design integrated with modified Hermite curves. We adopt a waveguide geometry of 800 nm × 300 nm in the simulations, with the material refractive index derived from the LPCVD process. Simulations conducted at a wavelength of 1311 nm reveal that the proposed bends exhibit bending losses of 0.076 dB, 0.021 dB, 0.0055 dB, and 0.0012 dB per 90° bend, corresponding to bending radii of 15 μm, 20 μm, 25 μm, and 30 μm, respectively. Furthermore, the fabrication tolerance and wavelength dependence of the proposed design are systematically investigated to verify its practical applicability.

Keywords: waveguide bend; silicon nitride; modified Hermite curve

1. Introduction

Silicon nitride (SiN) has emerged as a highly promising material for photonic applications, owing to its broad optical transparency window, ultra-low propagation loss, moderate Kerr nonlinearity, excellent high-power handling capability, superior thermal stability, and compatibility with the complementary metal-oxide-semiconductor (CMOS) process [1,2]. SiN exhibits negligible two-photon absorption at communication wavelengths—a nonlinear effect that can act as a limiting factor in silicon optical waveguides. This property enables SiN devices to handle higher optical power, establishing it as the preferred platform for nonlinear applications such as frequency comb generation [2]. Unlike silicon, which becomes opaque at wavelengths below ~1.1 μm, SiN maintains high optical transparency over an exceptionally broad wavelength range, spanning from the visible spectrum to the mid-infrared region. This unique characteristic renders it an ideal platform for applications beyond traditional communication bands, such as displays, biophotonics, mid-infrared sensing [3].

SiN optical waveguides are renowned for their extremely low propagation loss, which now routinely falls within the decibel-per-meter (dB/m) range [4]. This performance is mainly attributed to the high-quality SiN thin films deposited via techniques such as LPCVD and the relatively low refractive index contrast between SiN (~2.0) and silicon

 https://doi.org/10.3390/photonics13020175

dioxide (~1.5)—an attribute that reduces scattering loss induced by sidewall roughness. Recent demonstrations have reported losses as low as 0.045 dB/m [5] and 0.034 dB/m [6].

However, the lower refractive index contrast of SiN compared to silicon-on-insulator (SOI) necessitates a larger bending radius, which in turn limits the integration density of SiN PICs [7]. Therefore, developing a design methodology for compact, low-loss bent waveguides is of great significance for advancing PIC implementations on the SiN platform.

To date, The O-band has become the preferred solution for short-reach optical communications, especially for applications in data centers [8]. For example, 1.6-Tb/s transmission over 10 km in the O-band using 400-Gb/s/lane spatial division multiplexing (SDM) channels has been demonstrated, highlighting its critical role in high-speed data transfer [8]. However, to the best of our knowledge, the minimum bending radius for the O-band SiN waveguide is 30 μm and the loss is 0.004 dB/90°, which has been realized through an adiabatic design [9]. As for the low-loss O-band SOI waveguide, the bending radius can be reduced to 5 μm with a loss of 0.06 dB/90° [10]. This significant disparity underscores the critical need to investigate the trade-offs between compactness and propagation loss in O-band SiN bent waveguides. Despite its importance, relatively little research has been conducted on SiN bent waveguides operating at the O-band.

In this paper, we first briefly analyze the bending loss mechanism and review the state-of-the-art of SiN bent waveguides. Then we propose a novel bending design approach to realize compact, low-loss O-band SiN waveguides with enhanced wavelength insensitivity. Numerical simulations show bending loss of 0.076 dB, 0.021 dB, 0.0055 dB, and 0.0012 dB per 90° at radii of 15 μm, 20 μm, 25 μm, and 30 μm, respectively. We compare their performance with circular, and other advanced bending techniques, and analyze the impact of fabrication process errors.

2. Low Loss SiN Waveguide Bend

2.1. The SiN Material

Currently, there are two mainstream techniques for depositing SiN thin films: low-pressure chemical vapor deposition (LPCVD) and plasma-enhanced chemical vapor deposition (PECVD). Both have been widely applied in practice, and each possesses distinct technical characteristics [1].

Silicon nitride thin films prepared via LPCVD exhibit a relatively fixed refractive index, which ensures high performance stability for optical devices based on it. Additionally, these films contain extremely low levels of hydrogen bonds, endowing them with excellent low-loss properties. However, LPCVD has significant limitations. For one thing, its deposition process requires a high-temperature environment (900~1100 °C), making it incompatible with the silicon photonics process platform with doping process for modulation and detection functions. For another, the high-purity silicon nitride thin films produced by LPCVD are prone to stress-related issues, and the film will crack when grown to a certain thickness [7]. Thus, high thickness (>400 nm) remains a core challenge for LPCVD-based silicon nitride waveguides. A solution to this problem is the adoption of the damascene process, which was proposed by the Swiss Federal Institute of Technology Lausanne (EPFL) [11]. This process embeds silicon nitride (SiN) into silica trenches by combining etching and chemical mechanical polishing (CMP) techniques, and it can effectively address the stress issue. However, it suffers from two main drawbacks: the thickness and width of the SiN waveguide are difficult to precisely control; in addition, the CMP process applied to large-area SiN regions results in a dishing effect, where the concentrated pressure exerted by the polishing head leads to over-polishing, causing a central depression of 10–20 nm [12].

In contrast, PECVD silicon nitride waveguides offer notable integration advantages. They can be seamlessly compatible with the silicon photonics platform, enabling the

https://doi.org/10.3390/photonics13020175

construction of multi-layer optical waveguide structures. Meanwhile, film stress can be effectively adjusted by regulating process parameters, allowing the thickness of PECVD silicon nitride waveguides to 800 nm [13]. Nevertheless, PECVD also has prominent shortcomings. The silicon nitride thin films it produces typically contain more hydrogen bonds, which induce substantial absorption loss in waveguides. This loss is difficult to further reduce through conventional methods. While high-temperature annealing can mitigate this loss by decreasing hydrogen bond content (when temperature conditions permit), the transmission loss of PECVD silicon nitride waveguides remains significantly higher than that of LPCVD-prepared counterparts [14].

2.2. *Bending Loss*

Bending loss in waveguides mainly includes propagation loss, radiation loss, and mode conversion loss [15,16]. The propagation loss originates from absorption and scattering caused by material properties, growth defects, and fabrication imperfections. The surface scattering becoming the dominant factor after optimizing waveguide growth quality [16]. Radiation loss is mainly caused by waveguide bending, which weakens the confinement of light, resulting in leakage into the cladding [17]. This can be alleviated by optimizing the design or increasing the bending radius. The mode conversion loss arises from mode field mismatch at the junction of straight and bent waveguides, especially due to distorted mode profiles caused by sudden curvature changes, and can be reduced by designing structures with gradually varying curvature [18].

2.3. *State-of-the-Art in SiN Bent Waveguides*

Here, we briefly review the state-of-the-art of bent waveguides. Table 1 summarizes the results of SiN waveguide bends.

In 2016, the IMEC team proposed a low-loss SiN bent waveguide for the visible band by introducing an offset at the interface between the bent and the straight waveguides—with the configuration illustrated in Figure 1a. This waveguide was fabricated via the PECVD process. A bending loss of 0.215 dB/90° was achieved at a radius of 5 μm, representing a reduction of 0.4 dB/90° compared to that of conventional bent waveguides [19].

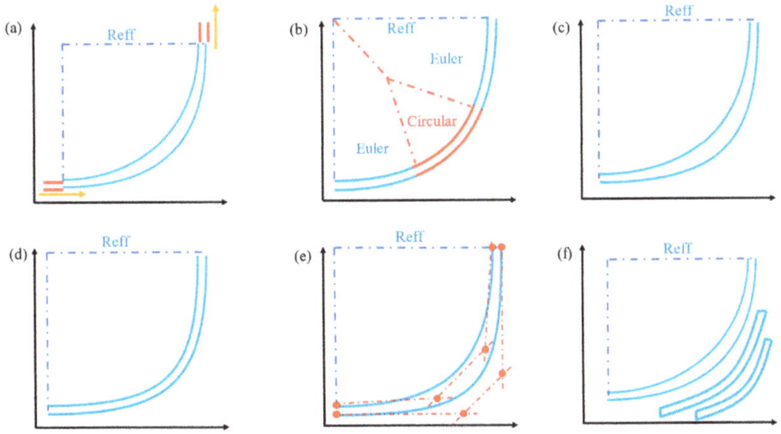

Figure 1. Schematic diagram of the selected SiN bent waveguide design method. (**a**) Bent waveguide with offset progressive contour. (**b**) Partial Euler waveguide bends. (**c**) Advanced bends. (**d**) Euler spiral. (**e**) Modified Bezier bent waveguide (**f**) Inverse-designed bends.

https://doi.org/10.3390/photonics13020175

In 2019, the team from Austrian Institute of Technology GmbH conducted a detailed analysis of various loss mechanisms of partially Euler-bent silicon nitride at a wavelength of 850 nm. They proposed a silicon nitride bent waveguide based on a partial Euler-bend geometry, as shown in Figure 1b. This structure optimized the transmission performance of small-radius optical waveguide bends and achieved a balance between the loss during the transition from the straight waveguide mode to the bent waveguide mode and the radiation loss of the bent waveguide mode. A bending loss of 0.007 dB/90° was achieved at a radius of 50 μm [20].

In 2020 and 2024, IMEC proposed a design method of advanced bends composed of width expansion and radius change at wavelengths of 850 nm and 1550 nm respectively, as shown in Figure 1c, and achieved extremely low bending losses through PECVD and LPCVD process methods respectively. Specifically, at a wavelength of 850 nm, the bending loss of a SiN waveguide with a radius of 15 μm decreases from 0.13 dB/90° for a conventional circular bends to 0.04 dB/90° [21]. And, at the wavelength of 1550 nm, the bending loss of a SiN waveguide bend with a radius of 25 μm was reduced from 0.048 dB/90° to 0.022 dB/90° [22].

In 2021, the team from Advanced Micro Foundry (AMF) proposed a so-called advanced waveguide bend based on the Euler spiral using a CMOS-compatible multi-layer PECVD SiN-on-SOI integration platform, as shown in Figure 1d. The curvature of this bend was designed to change linearly to reduce bending loss. The bending loss of SiN with a radius of 30 μm was reduced from 0.087 dB/90° for conventional circular bends to 0.037 dB/90° for the Euler bends [23].

In 2022, the team from Zhejiang University proposed a modified Bezier bend design for SiN PICs and validated its low-loss performance experimentally, with the configuration illustrated in Figure 1e. For a SiN bend with a bending radius of 12 μm, the measured bending loss was reduced to 0.32 dB per 90°—a significant improvement compared to 0.65 dB per 90° for conventional circular bends and 0.575 dB per 90° for traditional Bezier bends [24].

In 2025, the team from École de technologie supérieure (ÉTS) demonstrated the application of inverse design for realizing high-efficiency, compact bends in SiN waveguides, with a minimum bending radius of 6 μm (as illustrated in Figure 1f). These bends were specifically designed for the transverse electric (TE) polarization of single-mode waveguides. Devices were fabricated via electron-beam lithography (EBL), with the minimum feature size constrained to 160 nm to enhance compatibility with commercial CMOS foundries, and experimental results show that the free-form bends enabled by inverse design outperform conventional circular and Euler bends in terms of performance across four distinct footprint areas (ranging from 6 × 6 μm^2 to 21 × 21 μm^2). At bending radii of 6 μm, 11 μm, 16 μm, and 21 μm, the measured bending losses are 1.89 dB/90°, 0.18 dB/90°, 0.08 dB/90°, and 0.11 dB/90°, respectively [25].

In 2025, the TSMC team demonstrated an adiabatic SiN bent waveguide with PECVD and 65 nm process. The device exhibits a measured bending loss of 0.004 dB per 90° bend at 1311 nm, with a bending radius of 30 μm [9].

Table 1. State-of-the-art in compact SiN bends with experimental demonstrations.

Year/Ref.	Affiliation	Radius (μm)	Bending Loss (dB/90°)	Wavelength (nm)	Process	Waveguide Dimension (Width × Height)
2016/[19]	IMEC	3 4 5 7 10	1.403 0.633 0.215 0.054 0.034	638	PECVD	450 nm × 220 nm

https://doi.org/10.3390/photonics13020175

Table 1. *Cont.*

Year/Ref.	Affiliation	Radius (μm)	Bending Loss (dB/90°)	Wavelength (nm)	Process	Waveguide Dimension (Width × Height)
2019/[20]	AIT	50	0.007	850	NA	680 nm × 160 nm
2020/[21]	IMEC	5 10 15 20	1.82 0.25 0.04 0.01	850	PECVD	600 nm × 300 nm
2021/[23]	AMF	30 40 50	0.037 0.031 0.023	1550	PECVD SiN on SOI	1000 nm × 400 nm
2022/[24]	ZJU	8 10 12	1.225 0.525 0.32	1550	LPCVD	1200 nm × 400 nm
2024/[22]	IMEC	15 20 25	0.13 0.03 0.025	1550	LPCVD	1000 nm × 400 nm
2025/[25]	ÉTS	6 11 16 21	1.86 0.18 0.08 0.11	1550	PECVD	850 nm × 400 nm
2025/[9]	TSMC	30	0.004	1311	PECVD SiN on SOI	800 nm × NA

3. Device Structure and Design Method

3.1. Hermite Bend

Hermite curves find extensive applications in computer graphics and animation for defining the trajectories of moving objects—such as characters or vehicles—thereby ensuring a smooth and natural appearance to their motion [26].

Hermite curves are defined through Hermite interpolation, a method that ensures the continuity of higher-order derivatives at the connection points. This property is crucial for mitigating radiation loss and higher mode excitation in bent waveguides, thereby ensuring single-mode, low-loss transmission. A Hermite curve is parameterized by two endpoints and their corresponding tangent vectors. A schematic diagram of a Hermite curve bend, which is a third-order interpolation with two joints, is shown in Figure 2a. The expressions for the Hermite curve are as follows:

$$\begin{cases} c_1 = \left(1 - 3t^2 + 2t^3\right) \\ c_2 = \left(t - 2t^2 + t^3\right) \\ c_3 = \left(-t^2 + t^3\right) \\ c_4 = \left(3t^2 - 2t^3\right) \end{cases} \tag{1}$$

$$P_{(t)} = c_1 P_0 + c_2 M_0 + c_3 M_1 + c_4 P_1, \tag{2}$$

where P_0 and P_1 denote the starting and endpoints, respectively. While M_0 and M_1 represent their corresponding tangent vectors. The variable t is a parameter that traces the curve from the beginning (t = 0) to the end (t = 1). The Hermite curve is expressed as a linear combination of four basic functions and four weighting factors. As depicted in Figure 2a, a 90° Hermite bend can be formed by cascading two 45° Hermite bends, which implies that the parameters a_0 and a_1 are equal.

https://doi.org/10.3390/photonics13020175

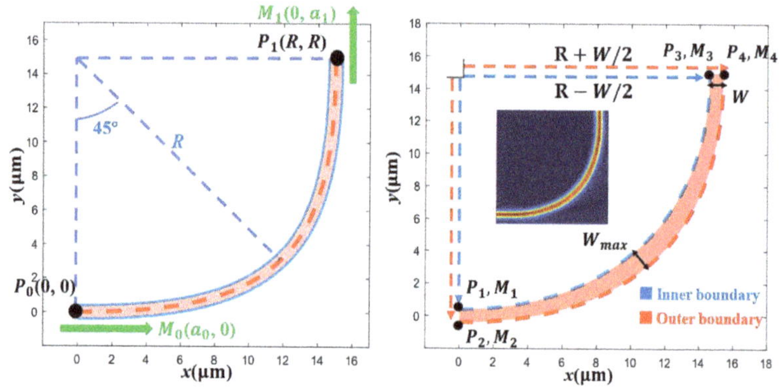

(a) (b)

Figure 2. (**a**) Hermite bend configuration and its design parameters. (**b**) Modified Hermite bend, design parameters and its electrical fields.

3.2. Modified Hermite Bend

We noticed that changing the width of a bent waveguide can effectively reduce losses [9,19,21,22,24,27]. This led us to the idea that by slowly changing the waveguide width based on the Hermite curve, the characteristics of bending losses can be improved. Upon its initial application in silicon photonic chips, the Hermite bend achieved exceptionally low losses and minimal mode excitation [28]. While Hermitian bends can effectively mitigate bending loss by adjusting tangential parameters, their performance is limited by a fixed width. To address this issue, improved Hermitian bends with variable widths can be constructed by introducing width variation. The graded-width design can effectively enhance the optical field confinement capability; an appropriate increase in width can raise the effective refractive index and mode confinement factor, thereby reducing the leakage of the optical field into the cladding [7,19].

For a specified bending radius R, the geometric configuration of the modified Hermite bend is constructed from two coupled Hermite curves rather than being defined by a single pair of start and end points (e.g., P_0 and P_1) for the entire waveguide path. Specifically, the two constituent Hermite curves take P_1 and P_2 as their respective starting points—corresponding to the two edge points at the initiation of the waveguide path—and P_3 and P_4 as their terminal points, which are the two edge points at the termination of the waveguide path. Notably, all these four key edge points (P_1, P_2, P_3, P_4) remain fixed for a given radius R. By employing the aforementioned method, two Hermite curves can be derived, as illustrated in Figure 2b, with their specific parameters detailed in Table 2. Here, W denotes the input waveguide width.

Table 2. Modified Hermite bend parameters.

Inner Boundary		Outer Boundary	
P	M	P	M
$P_1(W/2,\ 0)$	$M_1\ (a_1, 0)$	$P_2\ (-W/2,\ 0)$	$M_2\ (a_2, 0)$
$P_3\ (R - W/2,\ R)$	$M_3\ (0, a_3)$	$P_4\ (R + W/2,\ R)$	$M_4\ (0, a_4)$

The precise geometry of the modified Hermite bend is determined by the boundary lines of the two Hermite bends depicted in the figure. For the inner boundary of the modified Hermite bend, with a radius of $R - W/2$, the governing equation can be expressed as:

$$\begin{cases} x_{in} = c_1P_1(1) + c_2M_1(1) + c_3M_3(1) + c_4P_3(1) \\ y_{in} = c_1P_1(2) + c_2M_1(2) + c_3M_3(2) + c_4P_3(2) \end{cases} \quad (3)$$

For the modified Hermite bend outer boundary with a radius of $R + W/2$, the equation for the outer boundary can be expressed as:

$$\begin{cases} x_{out} = c_1P_2(1) + c_2M_2(1) + c_3M_4(1) + c_4P_4(1) \\ y_{out} = c_1P_2(2) + c_2M_2(2) + c_3M_4(2) + c_4P_4(2) \end{cases} \quad (4)$$

As depicted in Figure 2, the waveguide width varies, reaching its maximum at the 45° mark. Owing to the structural symmetry of the bent waveguide at 45°, the parameters are related such that $a_1 = a_3$ and $a_2 = a_4$. Consequently, for a given radius, the optimization process for achieving minimum loss is simplified to finding the optimal values for only two parameters, a_1 and a_2.

4. Simulation Results

We have designed 90° bending waveguides on a 300 nm LPCVD SiN platform with effective radii of 15 μm, 20 μm, 25 μm, and 30 μm, each having an input width of 0.8 μm. This specific width was chosen to ensure the waveguide operates in a single-mode regime. The platform includes a bottom buried oxide layer, a top silicon oxide cladding, and a silicon nitride core with a refractive index of 2.0.

For the modified Hermite bend structure, the optimal configuration corresponding to minimum bending loss can be achieved by adjusting parameters a_1 and a_2. To determine the ideal values of these two parameters, we conducted simulation analyses using the three-dimensional finite-difference time-domain (3D-FDTD) method targeting the quasi-fundamental transverse electric (quasi TE) mode, combined with the particle swarm optimization (PSO) algorithm. The two parameters a_1 and a_2 range from 15 to 80, with the key constraint condition during optimization set as $a_1 \leqq a_2$. A total of 100 iterations were performed in this optimization, and the iteration process is presented in the figure, with all iterations achieving convergence. The only figure of merit (FOM) for evaluating the bent structure is insertion loss (IL). The PSO results for the three bent structures with different radii are illustrated in Figure 3a.

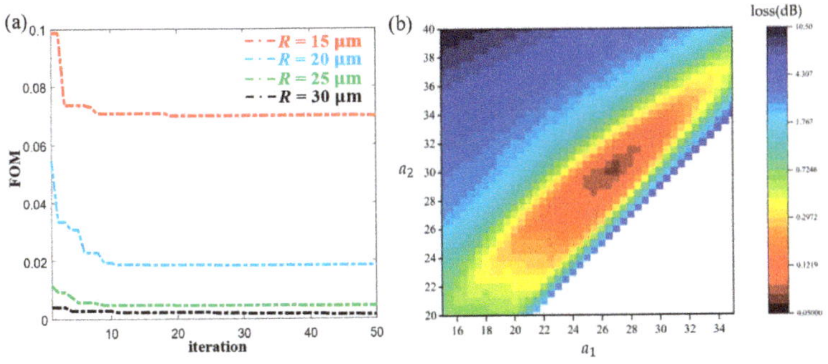

Figure 3. (**a**) PSO-based optimization of modified silicon Hermite bends for radii of 15 μm, 20 μm, 25 μm, and 30 μm. (**b**) 2D Heat map of Sweep Results of modified silicon Hermite bends for radii of 15 μm.

To verify the reliability of the PSO algorithm and intuitively demonstrate the parameter optimization process, we employed a parameter sweeping method for the auxiliary

https://doi.org/10.3390/photonics13020175

search of the optimal a_1 and a_2 values for the structure with a radius of 15 μm. The sweeping results are depicted as a two-dimensional (2D) heat map in Figure 3b. When a_2 is significantly smaller than a_1, the outer boundary of the bent structure intersects with the inner boundary, resulting in structural failure and a sharp increase in loss. Therefore, the results corresponding to these invalid parameters in the heat map have been excluded. As shown in Figure 3b, the sweeping results are in perfect agreement with the PSO results listed in Table 3, which further confirms the accuracy of the algorithm. The optimal parameter values of a_1 and a_2 for the modified Hermite bend, as well as the comparison results with the bending loss of the circular bend, are all summarized in Table 3, and the corresponding schematic diagrams of the structures are presented in Figure 4.

Table 3. Optimal values of bend parameters of modified Hermite bends and comparison of losses with circular bends.

R (μm)	Modified Hermite Bends			Circular Bend
	a_1	a_2	*IL* (dB/90°) @ 1311 nm	*IL* (dB/90°) @ 1311 nm
15	26.5	30	0.076	0.2
20	38.3	41.5	0.021	0.083
25	50.9	54.8	0.0055	0.048
30	63.3	66.9	0.0012	0.031

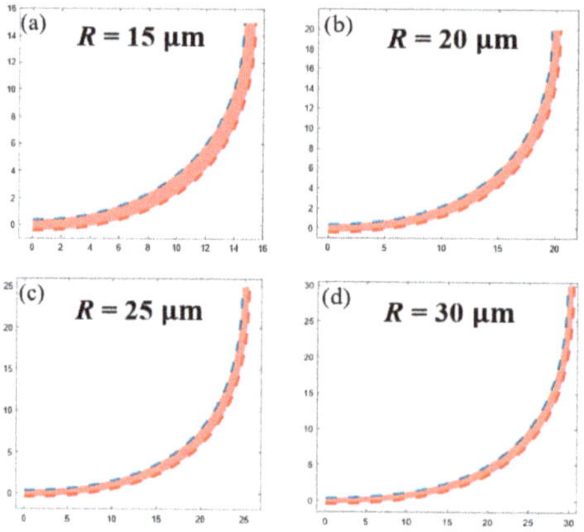

Figure 4. The structure of the modified Hermite bend when (**a**) *R* = 15 μm, (**b**) *R* = 20 μm, (**c**) *R* = 25 μm and (**d**) *R* = 30 μm.

To rigorously validate the performance of the proposed modified Hermite bend and ensure the fairness of the comparison, two sets of comparative simulations were conducted using 3D-FDTD under identical mesh precision and boundary settings. The first set benchmarks the proposed design against the conventional Hermite bend (both optimized via systematic scanning) and includes a comparison with the standard circular bend, with the results presented in Figure 5a. The second set contrasts the proposed structure with state-of-the-art alternatives—specifically the advanced bend [22] and the modified Bezier bend [24]—all of which were optimized using the same PSO algorithm to ensure algorithmic consistency; the corresponding spectral responses are illustrated in Figure 5b. For a quantitative assessment, the simulated insertion losses of the aforementioned bend structures at 1311 nm are summarized in Table 4. As clearly demonstrated, the proposed modified Hermite bend exhibits the lowest loss performance among all evaluated schemes. Furthermore, this design demonstrates superior wavelength insensitivity, indicating robust broadband potential for practical applications.

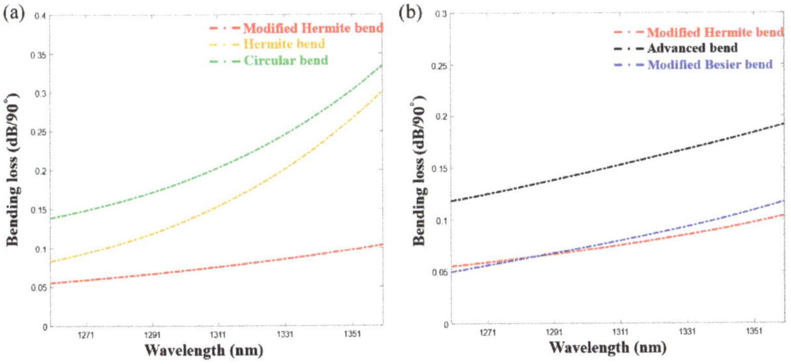

Figure 5. Simulated bending loss spectra for the TE0 mode at a bending radius of R = 15 µm. (**a**) Comparison of designs determined via systematic scanning: modified Hermite bend, conventional Hermite bend, and standard circular bend. (**b**) Comparison of designs optimized via the PSO algorithm: modified Hermite bend, advanced bend [22], and modified Bezier bend [24].

Table 4. Comparison of simulated insertion losses between the modified Hermite bend and other waveguide bend designs when R = 15 µm.

R = 15 µm	Structure				
	Circular Bend	Hermite Bend	Advanced Bend	Modified Bezier Bend	Modified Hermite Bend
IL (dB/90°) @ 1311 nm	0.2	0.161	0.151	0.083	0.076

Mathematically, cubic Hermite and cubic Bezier curves share an identical algebraic basis and are mutually convertible via a linear transformation matrix, as established in geometric modeling theory [29]. Consequently, they possess the same theoretical capacity for describing curve geometries. However, as evidenced in Figure 5b, the modified Hermite bend exhibits superior loss suppression and wavelength stability compared to its Bezier counterpart. This performance discrepancy is not intrinsic to the curves themselves but is attributed to the impact of parameterization efficiency on the heuristic optimization process.

The fundamental distinction lies in the definition of the geometry vector [29]: while the modified Bezier configuration requires four coordinate parameters to implicitly define the boundary derivatives, the modified Hermite formulation exploits the structural symmetry of the 90° bend to explicitly parameterize the tangent vector magnitudes. This reduces the search space dimensionality to merely two adjustable variables (a_1, a_2). By decoupling the tangent magnitudes from spatial coordinates, this reduction effectively mitigates the parameter redundancy and coupling inherent in the Bezier control polygon. Consequently, this streamlined parameterization minimizes the risk of the PSO algorithm stagnating in local optima and significantly accelerates convergence by lowering the computational overhead of 3D-FDTD iterations.

To evaluate the fabrication tolerance of the modified Hermite bend, we analyzed the influence of waveguide width and height variations of ±20 nm. We used 3D-FDTD simulations to evaluate the modified Hermite bend with a radius of 15 µm under varying width and height conditions. As illustrated in Figure 6a, our proposed design exhibits minimal loss variation across a 100 nm wavelength range under width deviations of ±20 nm. As shown in Figure 6b, it is observed that height has a more pronounced impact on the loss of the bent waveguide compared to width.

https://doi.org/10.3390/photonics13020175

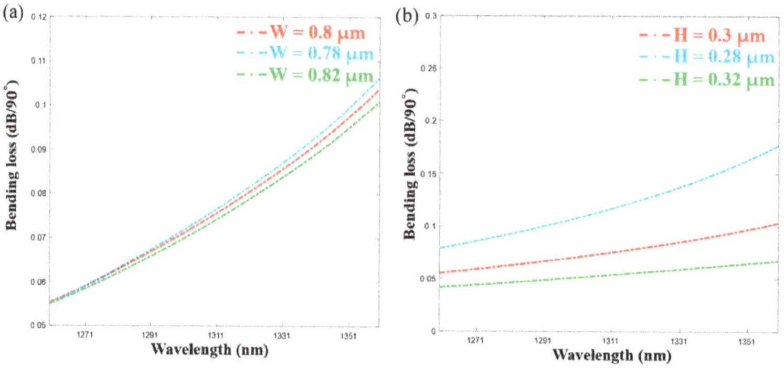

Figure 6. Simulated bending loss spectra of (**a**) Width (*W*) ± 20 nm (**b**) Height (*H*) ± 20 nm modified Hermite bends for *R* = 15 μm.

5. Discussion

The performance of the proposed modified Hermite bent waveguide is evaluated by comparison with state-of-the-art SiN bent waveguides, as summarized in Table 5. Most comparative designs are optimized for the C-band, while the proposed design focuses on the O-band—a critical wavelength range for short-reach data center communications [4]. Meanwhile, this design exhibits superior wavelength insensitivity, as shown in Figure 6a, and adopts a compact cross-section of 800 nm × 300 nm, which is more conducive to high-density integration. The graded-width Hermite curve design enhances the optical field confinement capability, illustrated in Figure 7, shifting the peak of the mode field away from the waveguide sidewalls and reducing scattering loss. Additionally, the design demonstrates robust fabrication tolerance: minimal loss fluctuation is observed under a waveguide width deviation of ±20 nm.

Table 5. Comparison of Simulated Losses for Different SiN Bent Waveguides.

Year/Ref.	Affiliation	Radius (μm)	Bending Loss (dB/90°)	Wavelength (nm)	Method	Waveguide Dimension (Width × Height)
2022/[24]	ZJU	8 10 12	0.95 0.41 0.23	1550	Modified Bezier bend	1200 nm × 400 nm
2024/[22]	IMEC	15 20 25	0.087 0.027 0.022	1550	Advanced bend	1000 nm × 400 nm
2024/[30]	UVA	15 20	~0.5 ~0.2	1550	Asymmetric Coupled Waveguides with Partial Euler Bends	1000 nm × 300 nm
2025/[25]	ETS	11 16 21	0.17 ~0.04 0.042	1550	Inverse-Designed	850 nm × 400 nm
This work	BUPT	15 20 25 30	0.076 0.021 0.0055 0.0012	1311	Modified Hermite Bend	800 nm × 300 nm

https://doi.org/10.3390/photonics13020175

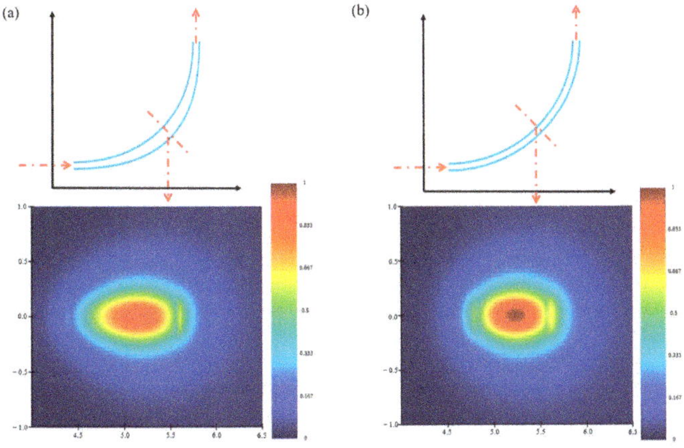

Figure 7. Simulated cross-sectional mode field intensity distributions at the 45° position for bent waveguides with R = 15 µm at 1311 nm. (**a**) Conventional circular bent waveguide; (**b**) Modified Hermite bent waveguide.

6. Conclusions

The proposed methodology facilitates the efficient design of small-radius, low-loss O-band SiN bends using only two parameters, which simultaneously reduces computational overhead. Fundamentally, this approach constitutes an interpolating polynomial, which could be further optimized in future work by configuring additional position coordinates and tangents. This methodology can also be extended to the structural design of a wider range of devices. Its potential for large-scale photonic integration establishes it as a significant contribution to the progress of this field.

In summary, we propose an efficient bent waveguide design methodology that employs a modified Hermite curve and the particle swarm optimization (PSO) algorithm to determine the geometry of the curve. Using this method, we design low-loss O-band SiN bends with radii of 15 µm, 20 µm, 25 µm, and 30 µm. The comparison with other bending design methods shows that the proposed method not only exhibits lower loss but also demonstrates superior wavelength insensitivity and process tolerance.

Author Contributions: Conceptualization, D.L. and L.Z.; methodology, D.L., E.Z., Y.X., L.Z., Y.Z. and X.Y.; software, D.L., E.Z. and X.Y.; validation, D.L., E.Z., Y.X. and Y.Z.; formal analysis, D.L., E.Z., Y.X. and L.Z.; investigation, D.L. and L.Z.; resources, D.L., L.Z. and X.Y.; data curation, D.L., Y.Z.; writing—original draft preparation, D.L.; writing—review and editing, D.L., E.Z., Y.X., L.Z., Y.Z. and X.Y.; visualization, D.L.; supervision, D.L., E.Z., Y.X., L.Z., Y.Z. and X.Y.; project administration, D.L., L.Z. and X.Y.; funding acquisition, L.Z., E.Z. and X.Y. All authors have read and agreed to the published version of the manuscript.

Funding: This work was supported in part by the Fund of State Key Laboratory of Information Photonics and Optical Communications (Beijing University of Posts and Telecommunications) (IPOC2024ZZ03), China, in part by the National Natural Science Foundation of China (U24B20135), in part by the BUPT Excellent Ph.D. Students Foundation (CX20242039), and in part by Beijing Natural Science Foundation (L259017).

Institutional Review Board Statement: Not applicable.

Informed Consent Statement: Not applicable.

Data Availability Statement: The original contributions presented in this study are included in the article. Further inquiries can be directed to the corresponding author.

Conflicts of Interest: The authors declare no conflicts of interest.

References

1. Blumenthal, D.J.; Heideman, R.; Geuzebroek, D.; Leinse, A.; Roeloffzen, C. Silicon Nitride in Silicon Photonics. *Proc. IEEE* **2018**, *106*, 2209–2231. [CrossRef]
2. Li, Z.; Fan, Z.; Zhou, J.; Cong, Q.; Zeng, X.; Zhang, Y.; Jia, L. Process Development of Low-Loss LPCVD Silicon Nitride Waveguides on 8-Inch Wafer. *Appl. Sci.* **2023**, *13*, 3660. [CrossRef]
3. Kim, E.; Carpenter, L.; Smith, A.M.; Tison, C.C.; Coleman, D.; Leake, G.; Fanto, M.; Rizzo, A.J.; García-Blanco, S.M.; Cheben, P. Ultra-low loss dispersion-engineered silicon nitride waveguides on 300 mm wafers. In Proceedings of the Integrated Optics: Devices, Materials, and Technologies XXIX, San Francisco, CA, USA, 27–30 January 2025. [CrossRef]
4. Chandran, S.; Bian, Y.; Lee, W.S.; Abumazwed, A.; Liu, N.; Xu, L.; Ding, H.; Aboketaf, A.; Rakowski, M.; Dezfulian, K.; et al. High performance silicon nitride passive optical components on monolithic silicon photonics platform. In Proceedings of the Optical Fiber Communication Conference, San Diego, CA, USA, 24–28 March 2024. [CrossRef]
5. Bauters, J.F.; Heck, M.J.R.; John, D.D.; Barton, J.S.; Bruinink, C.M.; Leinse, A.; Heideman, R.G.; Blumenthal, D.J.; Bowers, J.E. Planar waveguides with less than 0.1 dB/m propagation loss fabricated with wafer bonding. *Opt. Express* **2011**, *19*, 24090–24101. [CrossRef] [PubMed]
6. Liu, K.; Jin, N.; Cheng, H.; Chauhan, N.; Puckett, M.W.; Nelson, K.D.; Behunin, R.O.; Rakich, P.T.; Blumenthal, D.J. Ultralow 0.034 dB/m loss wafer-scale integrated photonics realizing 720 million Q and 380 μW threshold Brillouin lasing. *Opt. Lett.* **2022**, *47*, 1855–1858. [CrossRef] [PubMed]
7. Xiang, C.; Jin, W.; Bowers, J.E. Silicon nitride passive and active photonic integrated circuits: Trends and prospects. *Photon. Res.* **2022**, *10*, A82–A96. [CrossRef]
8. Taniguchi, H.; Nakamura, M.; Hamaoka, F.; Mori, T.; Shibahara, K.; Matsui, T. 1.6-Tb/s 10-km transmission in O-band using 400-Gb/s/lane SDM channels enhanced by trellis path-limitation MLSE. *J. Light. Technol.* **2024**, *42*, 4338–4346. [CrossRef]
9. Lu, H.-Y.; Tang, C.-T.; Chen, C.-H.; Tan, B.-X.; Liu, W.-K.; Chiang, H.-Y. Low-loss high-uniformity silicon nitride optical building blocks integrated on silicon photonics platform. In Proceedings of the 2025 Optical Fiber Communications Conference and Exhibition (OFC), San Francisco, CA, USA, 30 March–3 April 2025. Available online: https://ieeexplore.ieee.org/stamp/stamp.jsp?tp=&arnumber=11046774&isnumber=11046206 (accessed on 8 February 2026).
10. Yuan, Y.; Peng, Y.; Sorin, W.V.; Cheung, S.; Huang, Z.; Liang, D.; Fiorentino, M.; Beausoleil, R.G. A 5 × 200 gbps microring modulator silicon chip empowered by two-segment z-shape junctions. *Nat. Commun.* **2024**, *15*, 918. [CrossRef]
11. Ji, X.; Wang, R.N.; Liu, Y.; Riemensberger, J.; Qiu, Z.; Kippenberg, T.J. Efficient mass manufacturing of high-density, ultra-low-loss Si_3N_4 photonic integrated circuits. *Optica* **2024**, *11*, 1397–1407. [CrossRef]
12. Bose, D.; Harrington, M.W.; Ilicshenko, A.; Liu, K.; Wang, J.; Chauhan, N.; Newman, Z.L.; Blumenthal, D.J. Anneal-free ultra-low loss silicon nitride integrated photonics. *Light Sci. Appl.* **2024**, *13*, 156. [CrossRef]
13. Yang, C.; Pham, J. Characteristic Study of Silicon Nitride Films Deposited by LPCVD and PECVD. *Silicon* **2018**, *10*, 2561–2567. [CrossRef]
14. Rickman, A.G.; Reed, G.T.; Namavar, F. Silicon-on-insulator optical rib waveguide loss and mode characteristics. *J. Light. Technol.* **1994**, *12*, 1771–1776. [CrossRef]
15. Melloni, A.; Carniel, F.; Costa, R.; Martinelli, M. Determination of bend mode characteristics in dielectric waveguides. *J. Light. Technol.* **2001**, *19*, 571–577. [CrossRef]
16. Gao, F.; Qin, L.; Chen, Y.-Y.; Jia, P.; Chen, C.; Liang, L.; Chen, H.; Zhang, X.; Ning, Y.-Q. Research progress of bent waveguide and its applications. *Chin. Opt.* **2017**, *10*, 176–193. [CrossRef]
17. Collin, R.E. *Field Theory of Guided Waves*, 2nd ed.; Wiley-IEEE Press: New York, NY, USA, 1991; pp. 1–864. Available online: https://ieeexplore.ieee.org/book/5265444 (accessed on 8 February 2026).
18. Kitoh, T.; Takato, N.; Yasu, M.; Kawachi, M. Bending loss reduction in silica-based waveguides by using lateral offsets. *J. Light. Technol.* **1995**, *13*, 555–562. [CrossRef]
19. Song, J.H.; Kongnyuy, T.D.; Stassen, A.; Mukund, V.; Rottenberg, X. Adiabatically bent waveguides on silicon nitride photonics for compact and dense footprints. *IEEE Photonics Technol. Lett.* **2016**, *28*, 2164–2167. [CrossRef]
20. Vogelbacher, F.; Nevlacsil, S.; Sagmeister, M.; Kraft, J.; Unterrainer, K.; Hainberger, R. Analysis of silicon nitride partial Euler waveguide bends. *Opt. Express* **2019**, *27*, 31394–31406. [CrossRef]
21. Song, J.H.; Kongnyuy, T.D.; De Heyn, P.; Lardenois, S.; Jansen, R.; Rottenberg, X. Low-loss waveguide bends by advanced shape for photonic integrated circuits. *J. Light. Technol.* **2020**, *38*, 3273–3279. [CrossRef]
22. Song, J.H.; Kongnyuy, T.D.; Prost, M.; Bandyopadhyay, A.; Dwivedi, S.; Carbajal Altamirano, D. Advanced bends and micro ring resonators in silicon nitride photonic waveguides for C-band. *IEEE Photonics J.* **2024**, *16*, 1–5. [CrossRef]
23. Gao, F.; Xie, W.; Li, B.; Bu, X.; Song, A.; Luo, X. Low-loss and compact bends on multi-layer SiN-on-SOI platform for photonic integrated circuits. *IEEE Photonics Technol. Lett.* **2021**, *33*, 1131–1134. [CrossRef]

https://doi.org/10.3390/photonics13020175

24. Sun, T.; Xia, M. Low loss modified Bezier bend waveguide. *Opt. Express* **2022**, *30*, 10293–10305. [CrossRef]
25. Pita Ruiz, J.L.; Dalvand, N.; Ménard, M. Inverse-designed 90-degree silicon nitride bends for the C band. *J. Light. Technol.* **2025**, *43*, 5804–5810. [CrossRef]
26. Chen, X.; Li, Q.; Zheng, K.; Li, Y. Research and application of obstacle avoidance path planning based on Hermite curve interpolation. In Proceedings of the 2023 5th International Conference on Robotics, Intelligent Control and Artificial Intelligence (RICAI), Hangzhou, China, 1–3 December 2023. [CrossRef]
27. Zhang, L.; Wang, K.; Chen, J.; Ma, T.; Li, W.; An, J. Ultra-compact low-loss modified N-adjustable bends. *J. Light. Technol.* **2023**, *41*, 2792–2800. [CrossRef]
28. Tao, P.; Yan, B.; Bai, M.; Zhou, M.; Sun, C.; Qi, Y.; Wang, Y.; Han, Z.; Fan, Y.; Wang, Q. Ultrahigh Q SOI ring resonator with a Hermite or B-spline curved waveguide. *Appl. Opt.* **2024**, *63*, 6233–6239. [CrossRef]
29. Hughes, J.F.; van Dam, A.; McGuire, M.; Sklar, D.F.; Foley, J.D.; Feiner, S.K.; Akeley, K. Splines and Subdivision Curves. In *Computer Graphics: Principles and Practice*, 3rd rev. ed.; Hughes, J.F., van Dam, A., McGuire, M., Sklar, D.F., Foley, J.D., Feiner, S.K., Akeley, K., Eds.; Addison-Wesley: Reading, MA, USA, 2014; pp. 595–598.
30. Chamorro-Posada, P. Silicon Nitride Bent Asymmetric Coupled Waveguides with Partial Euler Bends. *Photonics* **2024**, *13*, 218. [CrossRef]

https://doi.org/10.3390/photonics13020175

photonics

MDPI

Article

PN Junction Optimization for High-Speed Silicon Photonic Modulators

Mahmoud Hamouda [1,2], Carine Mankarious [1,2,*], Aser El-Dahshan [1,2], Alaa Fathy [1,2], Eslam El-Fiky [3] and Diaa Khalil [1,2]

[1] Faculty of Engineering, Ain Shams University, Cairo 11517, Egypt
[2] InfiniLink, Cairo 4732030, Egypt
[3] Electrical Engineering Department, Alexandria University, Alexandria 5423021, Egypt
* Correspondence: carinemankarious@outlook.com

Abstract

PN-junction-based modulators are widely used in silicon photonic transceivers for different applications. Different junction shapes have been proposed in the literature. This work studies the optimization of the PN junction by tailoring the doping profile to achieve a high-efficiency modulator with sufficient bandwidth. For this purpose, a new N-shaped junction is proposed, which achieves superior performance compared to other junction shapes. The proposed junction has an efficiency that is 60% better than that of the lateral PN junction for the same doping condition, while maintaining a high bandwidth similar to other junctions such as the L-shaped and S-shaped designs. A junction design with an estimated RC bandwidth between 70 GHz and 94 GHz is also proposed. The impact of using the proposed junction in micro-ring modulators (MRMs) is also studied. N-shaped junctions in MRM demonstrated a 112% increase in electro-optic bandwidth over the vertical PN junction, with 60% and 140% improvements in extinction ratio (ER) and optical modulation amplitude (OMA), respectively, compared to the lateral PN junction.

Keywords: silicon photonics; ring modulator; PN junctions; data center interconnects

1. Introduction

Over the past decade, silicon photonics (SiPh) has received considerable recognition due to its low cost, high performance, and CMOS fabrication compatibility [1]. The demand for ultra-high-speed networks has grown significantly, leading to silicon photonics attracting substantial attention, particularly in the fabrication of optical transceivers for both long-haul and short-reach optical communication links.

For high-speed networks, fast optical phase shifters are essential components in photonic integrated circuits (PICs). The landscape of optical modulators is vast, with modulation schemes depending on phase or amplitude modulation. Existing literature reports the implementation of optical modulators on multiple platforms [2]: silicon [3], monolithic barium titanate (BTO) on silicon [4], thin-film lithium niobate (TFLN) on silicon [5], etc., each of which offers different material properties, leading to the exploration of a plethora of operating principles such as the plasma dispersion effect [6], the Pockels effect [7], the Franz–Keldysh (FK) effect [8], and the quantum-confined Stark (QCS) effect [9].

With the dense, cost-effective, and power-efficient integration of photonics on SiPh platforms [10–12], the deployment of the plasma dispersion effect in high-speed modulators has become the most widely adopted. The plasma dispersion effect depends essentially on the creation of a refractive index change in a silicon waveguide as a result of the movement of the carriers in the formed PN junction through injection, accumulation, or depletion of charges [12]. The junction performance as a modulator or phase shifter is usually evaluated by the applied voltage required to achieve a phase shift of π (V_π) through a waveguide of length L_π. The reduction in the value of the product $V_\pi L_\pi$ is highly desired to achieve a high-speed modulator with low voltage swing and a small footprint, as an increase in the modulator length directly reduces its bandwidth. This product for injection and accumulation junctions can be as low as 0.16 V.cm [11] and even 0.058 V.cm [13], while depletion junctions can reach data rates as high as 100 Gbps (OOK) [14]. The work presented in this paper focuses on plasma-dispersion-based modulators using the depletion principle due to the high speeds they can achieve as a result of their limited capacitance compared to injection- and accumulation-based junctions.

By increasing the overlap of the formed depletion region with the optical mode, stronger spatial variation can be achieved. This means that the change in the free carrier density results in a stronger change in the material refractive index, which translates into an enhancement in modulation efficiency.

A lot of research has been conducted in this direction by altering the junction doping profile, some of which includes U-shaped junctions [15], zigzag junctions [16], and interleaved junctions [17]. All of the previous work shows a clear trade-off between enhancing modulation efficiency and maintaining optical losses and capacitance.

This paper describes how to obtain the required junction and how the resulting junctions perform so they can accordingly serve different applications based on the previously mentioned trade-offs. The rest of this paper is organized as follows: in Section 2, the theory behind optical modulation is thoroughly explained, and the method of optimization is described in sufficient detail. The simulation results are presented in Section 3, along with the subsequent discussion in Section 4. Finally, Section 5 summarizes all the findings of this study and their relevance.

2. Theory and Methods

Within the plasma dispersion effect, changes in carrier concentration under an applied electric field lead to variations in silicon absorption. Material absorption is generally described by the imaginary part of the refractive index, and according to the Kramers–Kronig relation, the variation in the imaginary part of the complex refractive index is accompanied by a corresponding change in its real part. Soref and Bennet [18] calculated the changes in real (Δn) and imaginary ($\Delta\alpha$) parts of the refractive index due to changes in carrier densities in silicon at a wavelength of 1.55 μm to be [18]:

$$\Delta n = -\left[8.8 \times 10^{-22} \times \Delta n_e + 8.5 \times 10^{-18} \times (\Delta n_h)^{0.8}\right] \quad (1)$$

$$\Delta\alpha = 8.5 \times 10^{-18} \times \Delta n_e + 6.0 \times 10^{-18} \times \Delta n_h \quad (2)$$

where Δn_e and Δn_h are the changes in the electron and hole concentrations, respectively. The main design specifications for optical modulators are the modulation bandwidth (speed) and the modulation efficiency. Lateral PN (LPN) junctions [19], illustrated in Figure 1a, have played a significant role in optical modulation over the past decade due to their relatively simple fabrication (photolithography and doping), and high modulation speed resulting from their reduced junction capacitance. However, lateral junctions have limited depletion regions, and hence limited overlap with the optical mode in the structure, which results in low modulation efficiency. Another well-known junction is the vertical PN (VPN) junction [20], shown in Figure 1b, which exhibits high modulation efficiency with limited bandwidth due to its high capacitance.

(**a**) (**b**)

Figure 1. Waveguide cross-section view of (**a**) LPN junction and (**b**) VPN junction.

The figure of merit (FOM) typically used to characterize a modulator PN junction is $V_\pi L_\pi RC$, where $V_\pi L_\pi$ expresses the modulation efficiency, R is the junction resistance, and C is its capacitance. The RC value usually has a dominant contribution to minimizing the FOM, often resulting in an LPN-like junction. To mitigate this shortcoming, we penalize $V_\pi L_\pi$ over RC, which instead provides a junction profile with a more appropriate efficiency–bandwidth trade-off.

In this work, we consider a silicon rib waveguide with dimensions of 220 nm in height, 500 nm in width, and a 95 nm-high slab, with degenerately doped regions (N++ and P++) of concentration 10^{20} cm^{-3}, heavily doped regions (N+ and P+) of concentration 10^{19} cm^{-3}, and lightly doped regions (N and P) of peak concentrations 5×10^{18} cm^{-3} [21]. The junction is implemented using four doping masks (P-dope mask, shallow P-dope mask, semi-deep P-dope mask, and N-dope mask). Their design parameters are W1, W2, H-left, and H-right, and they are mapped as shown in Figure 2.

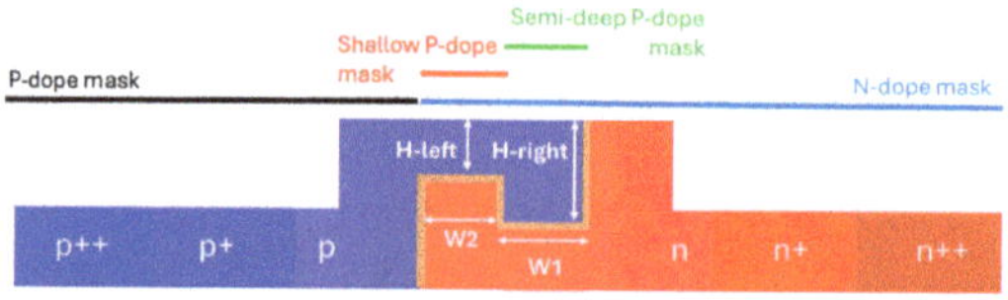

Figure 2. Doping masks used and junction parameters.

Using Ansys Lumerical CHARGE and FDE simulations, the PN-junction charge distribution is calculated, and the mode profile is examined. The junction is assumed to have a Pearson-4 distribution to approximate the real post-fabrication junction as closely as possible [22]. Comprehensive steady-state CHARGE simulations have been conducted to optimize the profile geometry, where the dimensions of the doping masks are swept to include all possible widths, offset from the center, and heights at a constant peak doping concentration. For each junction profile, the variation in the effective refractive index, the junction losses, resistance, and capacitance are extracted. The optimized junction parameters are then plugged into Lumerical INTERCONNECT in a microring modulator (MRM) to perform a clear comparison of the optical modulation amplitude (OMA), extinction ratio (ER), ratio of level mismatch (RLM), and electro-optic (EO) bandwidth, demonstrating the extent of the improvement and the different trade-offs compared to the conventional junctions. While we have chosen to show the performance based on an MRM, the junction can be used in different modulator designs such as a Mach–Zehnder modulator (MZM) or a ring-assisted Mach–Zehnder modulator (RAMZI).

3. Simulation Results

The results can be divided into three main areas of optimization: (1) modulation efficiency, (2) bandwidth, and (3) FOM with penalized $V_\pi L_\pi$. It is prudent to consider imposing constraints on physical design parameters, specified by the doping mask overlay and dimensions included in the aforementioned sweep, for the resulting junction to be compliant with the fabrication process limitations. Similar considerations apply to resulting optical parameters, where the previously discussed trade-offs should be maintained within acceptable ranges.

For each appropriate FOM value, the exact values of efficiency, bandwidth, and losses are evaluated. Figure 3 presents the junction modulation efficiency ($V_\pi L_\pi$) and RC bandwidth along with their associated losses, as shown in Figure 3a, and their relation to the penalized FOM ($(V_\pi L_\pi)^2$. RC), as shown in Figure 3b. Each point represents a junction with acceptable performance, where the point's shape can be mapped to the exact doping masks dimensions and placement. The results reveal the inherent trade-off between RC bandwidth and modulation efficiency, indicating that configurations with low penalized FOM tend to yield a denser cluster of points with higher efficiency but correspondingly lower bandwidth. The most efficient region, characterized by $V_\pi L_\pi \approx 0.4$ V.cm–0.5 V.cm and penalized FOM near 1 (V.cm)^2.pF.Ω, corresponds to bandwidths of 20 GHz–80 GHz. Conversely, designs achieving the highest bandwidth (~100 GHz) exhibit lower efficiency, with $V_\pi L_\pi$ values of 0.8 V.cm–1.1 V.cm.

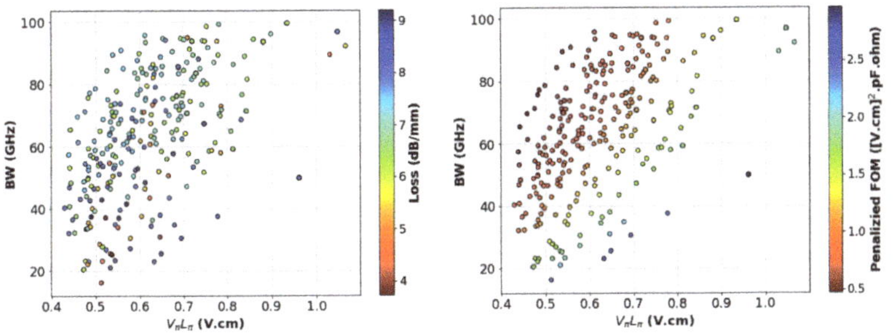

(a) (b)

Figure 3. Coarse sweep results showing trade-offs in junction modulation efficiency and bandwidth ($V_\pi L_\pi$ vs. BW), with (**a**) colored by loss and (**b**) colored by penalized FOM.

3.1. Optimized Junctions

The findings show that an L-shaped junction provides the highest efficiency at $V_\pi L_\pi = 0.439$ V.cm while still maintaining an appropriate bandwidth of almost 46.72 GHz. The charge profile of the junction is shown in Figure 4.

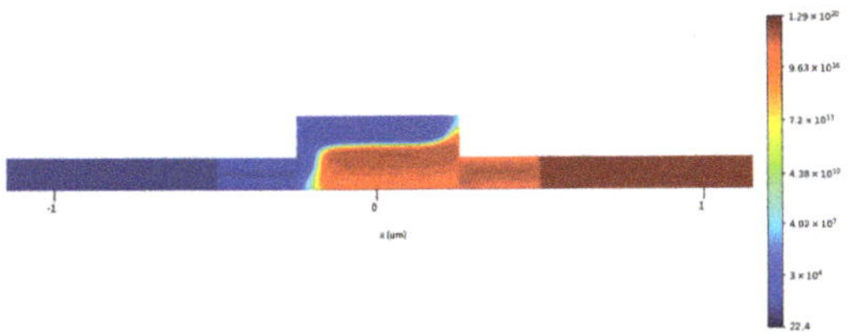

Figure 4. L-shaped junction profile.

On the other hand, when maximizing the bandwidth with a constraint placed on efficiency, where $V_\pi L_\pi$ is kept below 0.8 V.cm (a maximum 2× degradation compared to typical LPNs), the results shown in Figure 5 indicate that an S-shaped junction has the highest RC bandwidth at 103.35 GHz while maintaining an appropriate efficiency of almost 0.664 V.cm.

It can be deduced that a compromise between bandwidth and efficiency can manifest in a junction that combines both L-like and S-like profiles. When optimizing for the FOM with penalized $V_\pi L_\pi$, the resulting newly proposed junctions were found to have an N-like profile, as shown in Figure 6. It can be observed that different junction profiles have been achieved based on different doping mask alignments. Referring to the illustrative schematic in Figure 2, the mask alignments needed in the fabrication process to achieve the junctions shown below have the same shallow P-doping mask width (W2) of 125 nm. The difference in the doping profiles among those junctions lies in the achieved shallow and semi-deep P-doping region depths (H-left and H-right) and the semi-deep P-doping mask width (W1), with junctions N-shape I, II, and III having W1 of 125 nm, while N-shape IV has W1 of 62.5 nm. Furthermore, the doping energy by which the impurities are introduced into the silicon dictates the shallow- and semi-deep-doping region depths. The difference in the energy applied to each region gives the junctions their N-like profile. The previously mentioned N-shaped junctions have the same H-right of 155 nm and different H-left values of 110 nm, 155 nm, 65 nm, and 110 nm, respectively. Table 1 summarizes the geometrical parameters of the mentioned N-shaped junctions. Such geometries result in different modulation efficiencies and bandwidths due to the different depletion junction shapes. The bandwidth and efficiency for these junctions range from 69 GHz to 93 GHz and 0.54 V.cm to 0.61 V.cm, respectively. The different junction simulation results are summarized in Table 2. As mentioned previously, this FOM penalty adds emphasis to the modulation efficiency by increasing its weight; if it is desired to have a junction with higher bandwidth, with the efficiency being of lower priority but not as low as in junctions optimized solely for bandwidth, the weights can be tailored as needed.

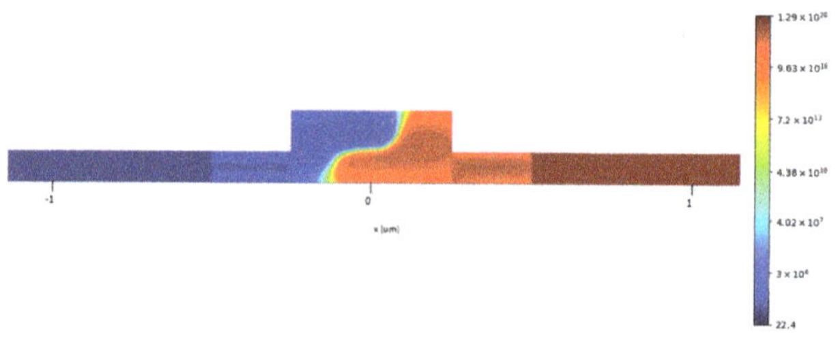

Figure 5. S-shaped junction profile.

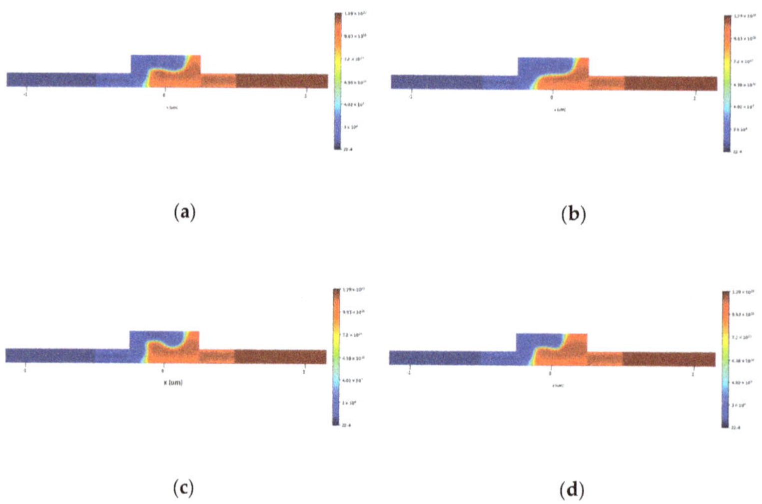

Figure 6. Optimized (**a**) N-shape I, (**b**) N-shape II, (**c**) N-shape III, and (**d**) N-shape IV junction profiles.

Table 1. Summary of the geometrical parameters for the chosen N-shaped junction configurations.

	N-Shape I	N-Shape II	N-Shape III	N-Shape IV
W1 (nm)	125	125	125	62.5
W2 (nm)	125	125	125	125
H-left (nm)	110	155	65	110
H-right (nm)	155	155	155	155

Table 2. Summary of the optimized junction results.

	LPN	VPN	L-Shape	S-Shape	N-Shape			
					I	II	III	IV
$V_\pi L_\pi$ (V.cm)	1.223	0.418	0.439	0.664	0.547	0.599	0.539	0.614
BW (GHz)	237.93	56.79	46.72	103.35	77.72	89.79	69.92	93.35

3.2. MRM Study

To demonstrate the significant impact of junction profile optimization on the performance of a typical compact modulator, this study uses an all-pass ring-based MRM as an example, as illustrated in Figure 7, with the following static transfer function [23]:

$$\frac{E_{out}}{E_{in}} = \frac{\sigma - \alpha e^{-j\beta L}}{1 - \sigma\alpha e^{-j\beta L}} \tag{3}$$

where α is the ring round-trip loss, κ^2 is the coefficient of the power coupled into the ring, σ^2 is the transmitted power coefficient, and L is the circumference of the ring. Meanwhile, the dynamic response can be expressed as follows [24]:

$$T(t) = \sigma(t) + \frac{\kappa(t)}{\kappa(t-\tau)}\alpha(t)e^{-j\phi_{\omega_0}(t)}[\sigma(t-\tau)T(t-\tau) - 1] \tag{4}$$

where τ is the round-trip time. The MRM used in this study has a radius of 10 µm. Using L-shaped and N-shaped junctions in the MRM, the ring is studied in Lumerical INTERCONNECT, where the RLM, OMA, ER, and EO bandwidths are examined. The RLM is the ratio between the smaller of the two modulation amplitudes and the larger one, assessing the symmetry of the modulation levels and therefore the modulation linearity. Performance improves as the RLM approaches unity, indicating perfect balance between the power levels and ideal linearity. The OMA is defined as $10\log(P_H - P_L)$, where P_H and P_L are the high and low power levels, respectively. Additionally, an OMA relative to 0 dBm input power (rOMA) is defined as $-10\log(P_H - P_L)$. Throughout the following sections, the rOMA is reported in its absolute value; accordingly, its reduction is desirable. The newly proposed junctions are compared to conventional ones, particularly the VPN and LPN junctions, as shown in Figure 8. For the junction comparison to be conclusive, a fixed peak-to-peak voltage swing of 2 V is applied, the same insertion loss (IL) of approximately 6 dB is enforced, as it lies near the maximum OMA region, and the same quality factor (Q) ranging from 3400 to 3500 is chosen for each design by varying the power coupling coefficient into the ring. Q is kept the same for all MRMs to assess the modulation efficiencies of the different PN junctions. After initial assessment of the MRM behavior at the previously mentioned junctions, it was found that increasing the doping concentration to 7×10^{18} cm^{-3} results in better overall performance for the LPN, L-shaped, and N-shaped junctions. Meanwhile, the VPN exhibits the highest round-trip losses; consequently, its doping level is kept at 5×10^{18} cm^{-3} for the MRM to demonstrate the same quality factor and insertion loss as its counterparts. Accordingly, the power coupling coefficient (κ^2) was set to 15%, 11%, 17%, and 17% for the LPN-, VPN-, L-shaped-, and N-shaped-junction-based MRMs, respectively. Table 3 summarizes the modulation efficiency, RC bandwidth, round-trip losses, and doping concentrations for each junction. The EO bandwidth of the MRM is limited by both the optical and the electrical (RC) bandwidths, where the dominant factor depends on the operating region. Accordingly, the RC bandwidth values provided in the table are not the only metric determining which junction can provide a higher 3 dB EO bandwidth, as explained in the next section.

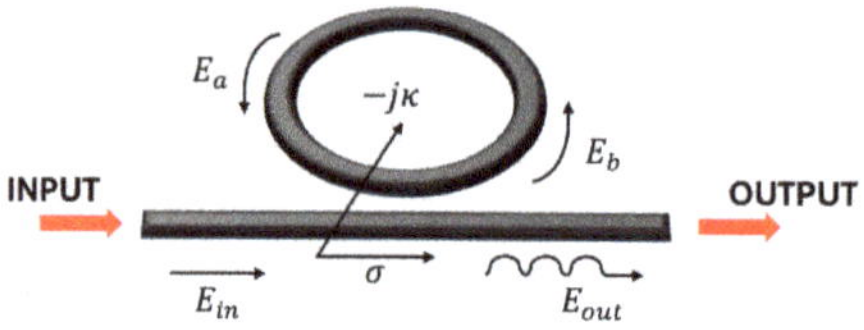

Figure 7. All-pass ring resonator.

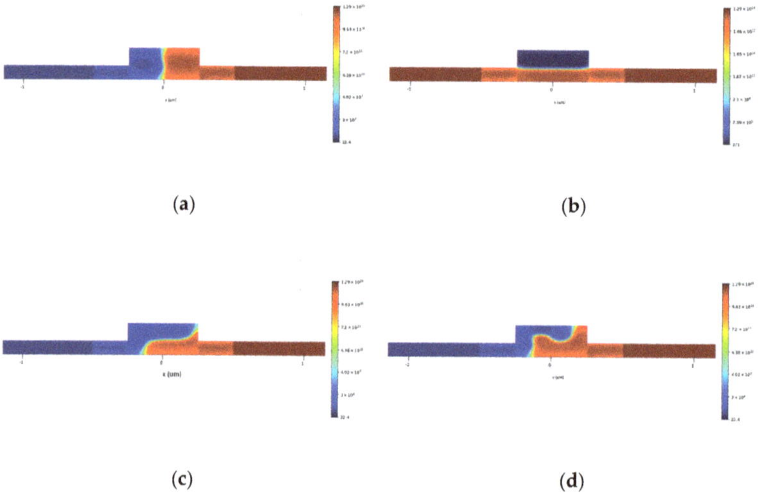

Figure 8. (**a**) LPN, (**b**) VPN, (**c**) L-shaped, (**d**) N-shaped junction profiles.

Table 3. Chosen junctions for MRM implementation.

	LPN	VPN	L-Shaped	N-Shaped
$V_\pi L_\pi$ (V.cm)	1.133	0.418	0.489	0.461
RC BW (GHz)	242	56.79	74	68.22
Loss (dB/round trip)	0.546	0.883	0.545	0.523
Doping concentration (cm^{-3})	7×10^{18}	5×10^{18}	7×10^{18}	7×10^{18}

- **Lateral (LPN) Junction**

As discussed previously, lateral junctions offer a smaller depletion region and therefore have minimal overlap between the free carriers and the optical mode. Implementing them in an MRM is expected to provide a wide bandwidth (low capacitance) at a low modulation efficiency. The MRM with the LPN junction shows a relatively weak resonance shift of 11 pm/V, as shown in Figure 9a. Although the LPN-MRM, among all the presented junctions, offers the widest 3 dB EO bandwidth at 80 GHz, as illustrated in Figure 10, it demonstrates the worst rOMA at 8.03 dBm, the lowest ER at 3.15 dB, and moderate linearity with an RLM of 77.16%.

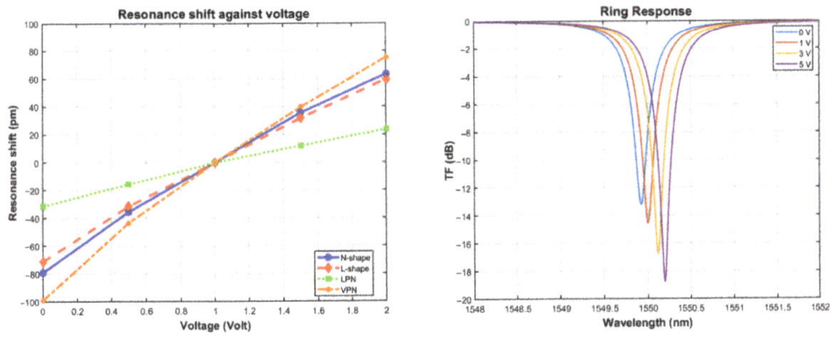

(a) (b)

Figure 9. (**a**) Resonance shift with the applied voltage for the MRMs with the proposed junctions. (**b**) Ring response for the N-shaped-junction-based MRM.

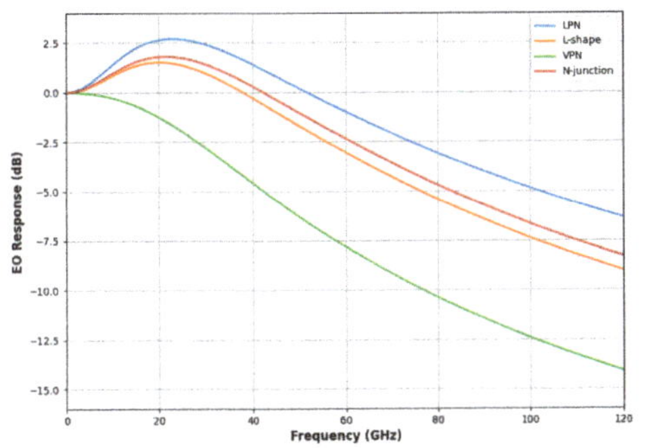

Figure 10. MRM EO response for the MRM with the proposed junctions.

- **Vertical (VPN) Junction**

In contrast, vertical junctions offer a relatively high overlap between the free carriers and the optical mode, making them the conventionally optimal choice for high modulation efficiency. The recorded resonance shift is significantly higher (3.64× increase) than that of the LPN, at almost 40 pm/V compared to 11 pm/V. Additionally, the VPN-MRM shows a high ER of 6.99 dB and a concurrent improvement in rOMA, compared to the LPN, decreasing to 4.96 dBm. Nevertheless, VPN junctions typically have limited RC bandwidth due to their large capacitance. Figure 10 shows a 3 dB EO bandwidth of 31 GHz, the lowest value observed among all the evaluated junctions. Figure 9a also shows that the VPN-MRM exhibits the poorest linearity, with an RLM of 70%.

- **L-shaped Junction**

L-shaped junctions offer a solution to the bandwidth limitation of VPNs while maintaining only a slight decrease in their high efficiency, due to the nature of their profile being a combination of both vertical and horizontal components, favoring the horizontal [25]. Accordingly, the MRM exhibits a resonance shift of approximately 30 pm/V and a 3 dB EO bandwidth of 60 GHz, as shown in Figure 10, while maintaining a high ER of 6.83 dB, a low rOMA of 5.17 dBm, and a significant enhancement in the RLM, compared to both the LPN- and VPN-MRMs, reaching 90.1%.

- **N-shaped Junction**

Extending the previously explained trade-off offered by the L-shaped junction, it can be observed that by alternating between vertical and lateral profiles along the junction, an optimal junction can be obtained. The predominantly N-shaped profile that appears in almost all junctions optimized for the $V_{\pi}L_{\pi}$-penalized FOM offers the best trade-off between efficiency, bandwidth, and junction losses. The proposed N-shaped junction, when implemented in the MRM at the same quality factor, exhibits a slightly higher resonance shift than the L-shaped junction at approximately 35 pm/V and the second-highest 3 dB EO bandwidth among the junctions under comparison at 66 GHz, as shown in Figure 10. Additionally, it exhibits the highest ER at 7.58 dB and high linearity with an RLM of 89.69%, simultaneously achieving the best rOMA at 4.9 dBm. This illustrates how reliable this profile approach is for increasing the bandwidth compared to the VPN while maintaining a robust enhancement in modulation efficiency compared to the LPN. Figure 9b shows the MRM response at different biasing voltages.

4. Discussion

4.1. Mode Overlap Comparison

VPN junctions tend to provide higher efficiency due to their horizontal profile that maximizes the overlap with the optical mode (28.1%), whereas the LPN exhibits the least overlap at 25.6%. On the other hand, the L-shaped junctions approach the lateral shape on the sides and the VPN shape in the middle, increasing the overlap compared to the LPN, with an overlap of 28.4%. Yet, it has a smaller depletion area, minimizing its capacitance, and hence increasing its RC bandwidth as illustrated in Section 3. Meanwhile, the proposed N-shaped junction provides a rational compromise between the two junctions; with the central section of the depletion region combining both horizontal and vertical profiles and the side sections being vertical, increasing the overlap to 31%, which is the highest among all the other discussed shapes, and simultaneously decreasing the depletion area, consequently decreasing the capacitance and increasing the RC bandwidth compared to the VPN. Intuitively, efficiency should increase with the overlap, but the efficiency also depends on the doping type with which most of the overlap occurs, and the depletion width variation with modulation. Figure 11 illustrates the differences discussed in the modal overlap and the electrostatic field distribution among the different junctions.

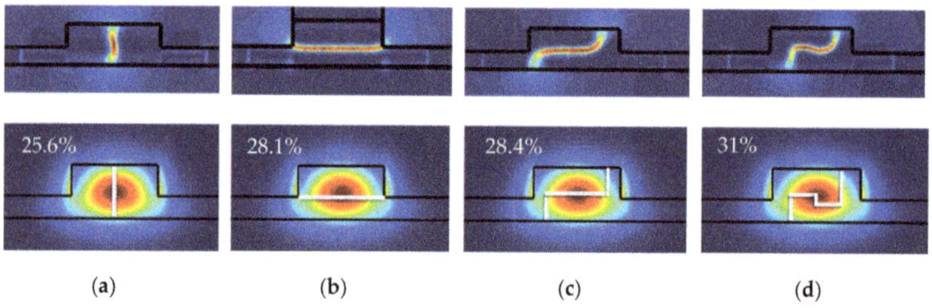

Figure 11. Mode overlap illustration. The first row is the electrostatic field distribution and the second is the charge profile of (**a**) LPN, (**b**) VPN, (**c**) L-shaped, and (**d**) N-shaped junctions, where the white lines represent the depletion region.

4.2. MRM Performance Comparison

As mentioned in Section 3.2., to perform an equitable assessment of the different junctions' implementation in the MRM, the applied voltage swing was chosen to be 2 V for the MRM to have an acceptable performance and concurrently remain CMOS-compatible. To clearly illustrate the effect of the used PN junction on the bandwidth of the MRM,

Figure 10 shows the different EO responses. Figure 12 depicts the 100 Gbaud PAM4 eye diagrams for the LPN-, VPN-, L-shaped-, and N-shaped-based MRM. The VPN exhibits the lowest 3 dB bandwidth while the N-shaped one offers the highest. All of those junctions have different round-trip losses, with the N-shaped being the least lossy and the VPN being the most, hence the noticeable distinct peaking in each one of them. It is evident how the newly proposed N-shaped junctions elevate the bandwidth performance capabilities of the MRM.

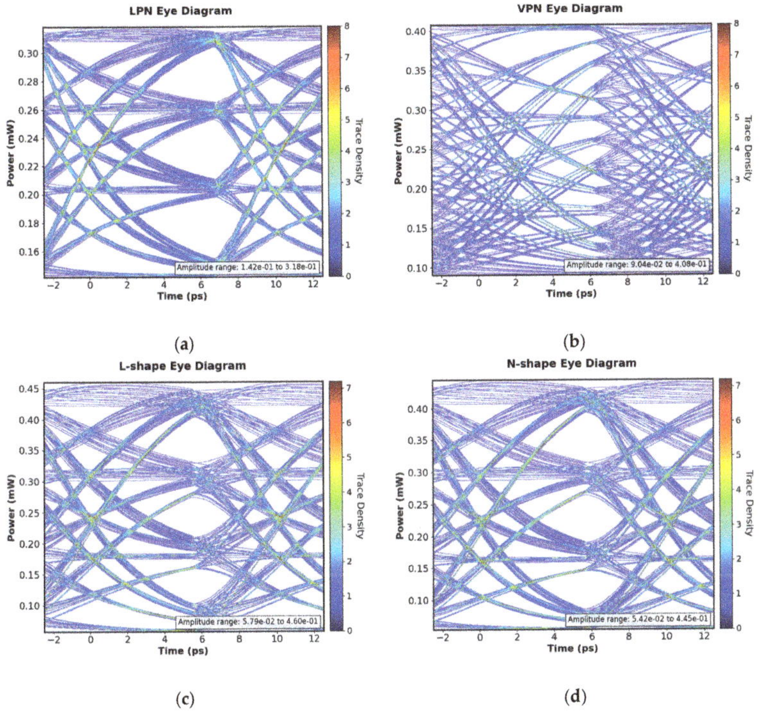

Figure 12. 100 Gbaud eye diagrams for the MRM with the (**a**) LPN, (**b**) VPN, (**c**) L-shaped, and (**d**) N-shaped junctions.

As discussed above when assessing the implementation of different junctions in the MRM, both the junction efficiency and the RLM are reflected in the resonance shift with the applied voltage as shown in Figure 9a, while the adequacy the EO bandwidth, or lack thereof, manifests in the eye opening as illustrated in Figure 12. The LPN exhibits the lowest resonance shift, hence the lowest efficiency. Even though the VPN demonstrates the highest resonance shift, it shows less ER and OMA than the N-shaped junction because the VPN losses are higher and its coupling coefficient is lower; thus, it is operating in the under-coupled region, with a higher dip in response compared to the N-shaped junction. The N-shaped junction bridges the gap between VPN and LPN most effectively, offering higher efficiency and comparable linearity with respect to the L-shaped counterpart.

Also, from the results presented in Figure 13, the LPN-MRM proves superior in terms of its EO bandwidth yet exhibits a relatively weak ER and the worst rOMA, while the VPN-MRM demonstrates a significant improvement in the rOMA and ER, with a moderate deterioration of the RLM compared to its lateral counterpart. The L-shaped junction retains the substantial improvement in the ER and rOMA achieved by the VPN, with only a slight decline. Notably, the N-shaped optimized junction offers the best trade-off among all the performance specifications as it minimizes the rOMA, simultaneously maximizing the ER and RLM, where ER, rOMA, and RLM are calculated based on the static MRM model. Table 4 summarizes the performance of the MRM for the various junction configurations presented above.

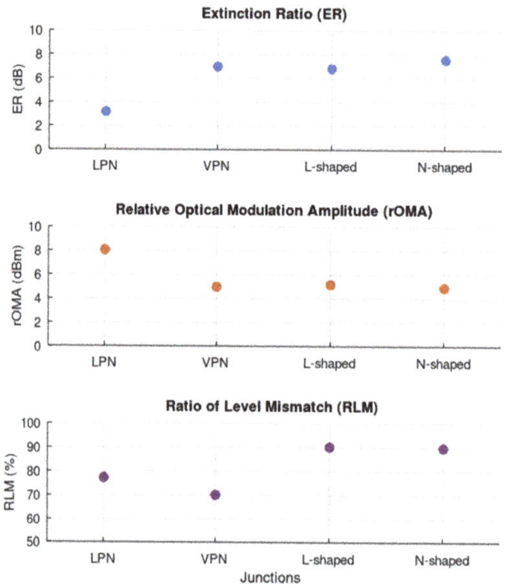

Figure 13. Summary of ER, rOMA, RLM, and 3 dB EO BW of different junctions.

Table 4. MRM performance summary at the different chosen junctions.

	LPN	VPN	L-Shaped	N-Shaped
3 dB BW (GHz)	80	31	60	66
rOMA (dBm)	8.03	4.96	5.17	4.9
ER (dB)	3.15	6.99	6.83	7.58
RLM (%)	77.16	70	90.1	89.69
Modal Overlap (%)	25.6	28.1	28.4	31

Furthermore, the impact of process variations on the N-shaped junction performance was investigated. The analyzed variations include deviations in the semi-deep P-doping mask alignment (±60 nm), the P-doping and N-doping mask alignments (±60 nm), the waveguide width (±20 nm), and the rib etch depth (±10 nm), representing realistic tolerances typically observed in standard SiPh fabrication processes. The efficiency ($V_\pi L_\pi$) varies between 0.45 V.cm and 0.67 V.cm, compared to the nominal value of 0.46 V.cm. Similarly, the loss fluctuates around the nominal 8.32 dB/mm, with a maximum deviation of approximately ±1 dB/mm. The bandwidth, on the other hand, exhibits the highest sensitivity to the process variations, ranging from 47.6 GHz to 107.5 GHz, with the nominal bandwidth being 68.22 GHz. Nevertheless, this bandwidth range can still provide adequate performance. These results confirm the robustness of the design under typical fabrication deviations. Table 5 details the impact of each process variation on the junction's performance parameters.

Table 5. Impact of process variations on N-shaped junction performance parameters.

Process Parameter		$V_{\pi}L_{\pi}$ (V.cm)	RC BW (GHz)	Loss (dB/mm)
Typical		0.461	68.22	8.32
Semi-deep mask alignment	−60 nm	0.474	57.5	9.36
	+60 nm	0.613	101.3	8.54
P/N-doping mask alignment	−60 nm	0.453	47.6	9.68
	+60 nm	0.674	107.5	7.88
Waveguide width	−20 nm	0.487	67.7	8.50
	+20 nm	0.497	71.7	8.50
Rib etch depth	−10 nm	0.466	64.2	8.44
	+10 nm	0.519	73.1	8.71

5. Conclusions

This work examined the potential of PN junction optimization for optical modulators, to support the requirements of different applications for high-speed optical communication networks. Our work examined the trade-off between modulation efficiency and RC bandwidth, and highlighted its impact on practical modulator (MRM) performance. It demonstrated the importance of tailoring the junction profile shape. The L-shaped junctions show an 80% improvement in efficiency compared to LPN but experience a significant compromise in the RC bandwidth. Meanwhile, when its performance is assessed in an MRM, it offered an adequate 3 dB EO bandwidth at 60 GHz, a high ER of 6.83 dB, and a low rOMA of 5.17 dBm. On the other hand, N-shaped junctions offer up to 60% improvement in efficiency, compared to LPN junctions, with a modest penalty to the RC bandwidth. When implemented in the MRM, they show a high ER of 7.58 dB, a low rOMA of 4.9 dBm, and a sufficiently fast modulation bandwidth of 66 GHz. The results and their related discussions show how the proposed N-shaped junctions offer the optimal middle ground among the presented trade-offs.

Author Contributions: Conceptualization, M.H., C.M. and A.E.-D.; methodology, M.H.; software, M.H. and A.E.-D.; validation, M.H., C.M. and A.E.-D.; formal analysis, M.H.; investigation, M.H., C.M. and A.E.-D.; resources, D.K., E.E.-F. and A.F.; data curation, M.H.; writing—original draft preparation, C.M.; writing—review and editing, M.H., C.M., A.E.-D., D.K., E.E.-F. and A.F.; visualization, M.H. and C.M.; supervision, D.K., E.E.-F. and A.F.; project administration, D.K., E.E.-F. and A.F. All authors have read and agreed to the published version of the manuscript.

Funding: This research received no external funding.

Data Availability Statement: The raw data supporting the conclusions of this article will be made available by the authors on request.

Conflicts of Interest: Mahmoud Hamouda, Carine Mankarious, Aser El-Dahshan, Alaa Fathy, and Diaa Khalil were employed by InfiniLink. The remaining author, Eslam El-Fiky, declares that the research was conducted in the absence of any commercial or financial relationships that could be construed as a potential conflict of interest.

References

1. Shekhar, S.; Bogaerts, W.; Chrostowski, L.; Bowers, J.E.; Hochberg, M.; Soref, R.; Shastri, B.J. Roadmapping the next Generation of Silicon Photonics. *Nat. Commun.* **2024**, *15*, 751. [CrossRef]
2. Rahim, A.; Hermans, A.; Wohlfeil, B.; Petousi, D.; Kuyken, B.; Thourhout, D.V.; Baets, R.G. Taking Silicon Photonics Modulators to a Higher Performance Level: State-of-the-Art and a Review of New Technologies. *Adv. Photonics* **2021**, *3*, 024003. [CrossRef]
3. Li, M.; Wang, L.; Li, X.; Xiao, X.; Yu, S. Silicon Intensity Mach–Zehnder Modulator for Single Lane 100 Gb/s Applications. *Photonics Res.* **2018**, *6*, 109–116. [CrossRef]

4. Eltes, F.; Mai, C.; Caimi, D.; Kroh, M.; Popoff, Y.; Winzer, G.; Petousi, D.; Lischke, S.; Ortmann, J.E.; Czornomaz, L.; et al. A BaTiO3-Based Electro-Optic Pockels Modulator Monolithically Integrated on an Advanced Silicon Photonics Platform. *J. Light. Technol.* **2019**, *37*, 1456–1462. [CrossRef]
5. Tang, Y.; Chen, Q.; Ma, X.; Liu, G.; Dai, X.; Lu, Q.; Guo, W. High Performance Thin-Film Lithium Niobate Modulator on Silicon Substrate with a Thick Silica Buffer Layer. *Opt. Express* **2025**, *33*, 20334–20344. [CrossRef]
6. Sun, J.; Kumar, R.; Sakib, M.; Driscoll, J.B.; Jayatilleka, H.; Rong, H. A 128 Gb/s PAM4 Silicon Microring Modulator with Integrated Thermo-Optic Resonance Tuning. *J. Light. Technol.* **2019**, *37*, 110–115. [CrossRef]
7. Koeber, S.; Palmer, R.; Lauermann, M.; Heni, W.; Elder, D.L.; Korn, D.; Woessner, M.; Alloatti, L.; Koenig, S.; Schindler, P.C.; et al. Femtojoule Electro-Optic Modulation Using a Silicon–Organic Hybrid Device. *Light Sci. Appl.* **2015**, *4*, e255. [CrossRef]
8. Verbist, J.; Verplaetse, M.; Lambrecht, J.; Srivinasan, S.A.; De Heyn, P.; De Keulenaer, T.; Pierco, R.; Vyncke, A.; Absil, P.; Yin, X.; et al. 100 Gb/s DAC-Less and DSP-Free Transmitters Using GeSi EAMs for Short-Reach Optical Interconnects. In Proceedings of the 2018 Optical Fiber Communications Conference and Exposition (OFC), San Diego, CA, USA, 11–15 March 2018; pp. 1–3.
9. Srinivasan, S.A.; Porret, C.; Vissers, E.; Favia, P.; De Coster, J.; Bender, H.; Loo, R.; Van Thourhout, D.; Van Campenhout, J.; Pantouvaki, M. High Absorption Contrast Quantum Confined Stark Effect in Ultra-Thin Ge/SiGe Quantum Well Stacks Grown on Si. *IEEE J. Quantum Electron.* **2020**, *56*, 5200207. [CrossRef]
10. Timurdogan, E.; Sorace-Agaskar, C.M.; Sun, J.; Shah Hosseini, E.; Biberman, A.; Watts, M.R. An Ultralow Power Athermal Silicon Modulator. *Nat. Commun.* **2014**, *5*, 4008. [CrossRef]
11. Ding, J.; Ji, R.; Zhang, L.; Yang, L. Electro-Optical Response Analysis of a 40 Gb/s Silicon Mach-Zehnder Optical Modulator. *J. Light. Technol.* **2013**, *31*, 2434–2440. [CrossRef]
12. Rahim, A.; Spuesens, T.; Baets, R.; Bogaerts, W. Open-Access Silicon Photonics: Current Status and Emerging Initiatives. *Proc. IEEE* **2018**, *106*, 2313–2330. [CrossRef]
13. Fujikata, J.; Takahashi, S.; Takahashi, M.; Noguchi, M.; Nakamura, T.; Arakawa, Y. High-Performance MOS-Capacitor-Type Si Optical Modulator and Surface-Illumination-Type Ge Photodetector for Optical Interconnection. *Jpn. J. Appl. Phys.* **2016**, *55*, 04EC01. [CrossRef]
14. Patel, D.; Veerasubramanian, V.; Ghosh, S.; Samani, A.; Zhong, Q.; Plant, D.V. High-Speed Compact Silicon Photonic Michelson Interferometric Modulator. *Opt. Express* **2014**, *22*, 26788–26802. [CrossRef] [PubMed]
15. Yong, Z.; Sacher, W.D.; Huang, Y.; Mikkelsen, J.C.; Yang, Y.; Luo, X.; Dumais, P.; Goodwill, D.; Bahrami, H.; Lo, P.G.-Q.; et al. U-Shaped PN Junctions for Efficient Silicon Mach-Zehnder and Microring Modulators in the O-Band. *Opt. Express* **2017**, *25*, 8425–8439. [CrossRef]
16. Xiao, X.; Li, X.; Xu, H.; Hu, Y.; Xiong, K.; Li, Z.; Chu, T.; Yu, J.; Yu, Y. 44-Gb/s Silicon Microring Modulators Based on Zigzag PN Junctions. *IEEE Photonics Technol. Lett.* **2012**, *24*, 1712–1714. [CrossRef]
17. Xiao, X.; Xu, H.; Li, X.; Hu, Y.; Xiong, K.; Li, Z.; Chu, T.; Yu, Y.; Yu, J. 25 Gbit/s Silicon Microring Modulator Based on Misalignment-Tolerant Interleaved PN Junctions. *Opt. Express* **2012**, *20*, 2507–2515. [CrossRef]
18. Soref, R.; Bennett, B. Electrooptical Effects in Silicon. *IEEE J. Quantum Electron.* **1987**, *23*, 123–129. [CrossRef]
19. Kim, Y.; Jin, T.; Bae, Y. A Comparative Simulation Study on Lateral and L-Shaped PN Junction Phase Shifters for Single-Drive 50 Gbps Lumped Mach–Zehnder Modulators. *Jpn. J. Appl. Phys.* **2021**, *60*, 052002. [CrossRef]
20. Maegami, Y.; Cong, G.; Ohno, M.; Okano, M.; Itoh, K.; Nishiyama, N.; Arai, S.; Yamada, K. High-Efficiency Silicon Mach-Zehnder Modulator with Vertical PN Junction Based on Fabrication-Friendly Strip-Loaded Waveguide. In Proceedings of the 2017 IEEE 14th International Conference on Group IV Photonics (GFP), Berlin, Germany, 23–25 August 2017; pp. 21–22. [CrossRef]
21. Witzens, J. High-Speed Silicon Photonics Modulators. *Proc. IEEE* **2018**, *106*, 2158–2182. [CrossRef]
22. Plummer, J.D.; Deal, M.D.; Griffin, P.B. *Silicon VLSI Technology: Fundamentals, Practice, and Modeling*; Prentice Hall Electronics and VLSI Series; Prentice Hall: Upper Saddle River, NJ, USA, 2000.
23. Bogaerts, W.; De Heyn, P.; Van Vaerenbergh, T.; De Vos, K.; Kumar Selvaraja, S.; Claes, T.; Dumon, P.; Bienstman, P.; Van Thourhout, D.; Baets, R. Silicon Microring Resonators. *Laser Photonics Rev.* **2012**, *6*, 47–73. [CrossRef]
24. Sacher, W.D.; Poon, J.K.S. Dynamics of Microring Resonator Modulators. *Opt. Express* **2008**, *16*, 15741–15753. [CrossRef]
25. Yuan, Y.; Peng, Y.; Sorin, W.V.; Cheung, S.; Huang, Z.; Liang, D.; Fiorentino, M.; Beausoleil, R.G. A 5 × 200 Gbps Microring Modulator Silicon Chip Empowered by Two-Segment Z-Shape Junctions. *Nat. Commun.* **2024**, *15*, 918. [CrossRef]

photonics

MDPI

Article

Ultra-High Spectral Contrast Nanobeam Photonic Crystal Cavity on Bending Waveguide

Ping Yu [1], Peihong Cheng [1,*], Zhuoyuan Wang [1], Jingrui Wang [1], Fangfang Ge [2], Huiye Qiu [3] and Daniel Kacik [4]

1 School of Electronic and Information Engineering, Ningbo University of Technology, Ningbo 315211, China; yuping@nbut.edu.cn (P.Y.); zywang@nbut.edu.cn (Z.W.); jruiwang@nbut.edu.cn (J.W.)
2 Ningbo Institute of Materials Technology and Engineering, Chinese Academy of Sciences, Ningbo 315201, China; gefangfang@nimte.ac.cn
3 Department of Mechanical and Electrical Engineering, Longyan University, Longyan 364012, China; qiuhuiye@lyun.edu.cn
4 Department of Physics, Faculty of Electrical Engineering, University of Zilina, 010 26 Zilina, Slovakia; daniel.kacik@uniza.sk
* Correspondence: chengpeihong@nbut.edu.cn

Abstract

In this article, one-dimensional photonic crystal cavities on bending waveguides (PCCoBW) used for achieving high-contrast spectra are proposed, analyzed, and experimentally verified on silicon on insulator (SOI). Both air and dielectric modes of the PCCoBW calculated by the finite-difference time-domain (FDTD) method show finger-ring-like mode profiles with the achievement of high-quality factors ($Q{\sim}10^6$), even when the bending radius is less than 50 times the lattice constant. Straight waveguides side-coupled to the cavity are used to access and measure mode resonances. The measured spectra show a high extinction ratio over 40 dB for dielectric modes and 20 dB for air modes, respectively. Both dielectric and air resonant modes are revealed with Q-factors over 3.3×10^4 and 7.9×10^4, respectively, for the coupled PCCoBWs. The proposed PCCoBW could be implemented as high-contrast notch filtering and would benefit a broad range of applications such as optical filters, modulators, sensors, or switches.

Keywords: photonic crystal; optical nanocavities; mode-dependent coupling; high extinction ratio

1. Introduction

By exploiting the design principles of photonic crystals [1–4], photonic crystal nanocavities can exhibit exceptional performance, characterized by an ultra-high quality factor ($Q > 10^6$) and an ultrasmall modal volume V, typically around $(\lambda/n)^3$. A one-dimensional photonic crystal cavity (or nanobeam photonic crystal cavity, NPCC) constructed on a periodic hole nanobeam not only has an ultra-high Q/V [5,6] but also renders an ultra-compact waveguide geometry suitable for high integration. The ultra-high Q/V metric means ultralow power consumption, ultrafast operation speed, and ultra-strong light–matter interaction [7]. On-chip applications, including lasers [8], filters [9], modulators [10], switches [11], and sensors [12,13], have been demonstrated on silicon on insulator (SOI), SiN, AlN, InP, GaAs, GaN, $LiNbO_3$, SiC, diamond, polymer, or chalcogenide. The fundamental cavity–waveguide configurations involved in these aforementioned NPCC applications are mainly classified into in-line coupling [14] and side-coupling [15,16]. Although finite-difference time-domain (FDTD) simulation predicts the high transmission of an optimized in-line coupling nanobeam photonic crystal cavity, the recorded experimental transmission is only −1.5 dB under ambient air-cladding [14]. For silica-encapsulated NPCCs, the reported insert loss is relatively high, with a typical value being under −10 dB [17].

Side-coupled NPCCs, however, provide high baseline transmission for non-resonant waves through the output port while sustaining high Q. They demonstrate attractive capabilities, such as serial integration, add/drop filtering [18], and multiplexing [12,13], which are not readily attainable from in-line devices. However, due to the inherent geometry limitation of straight nanobeams, the parallel side-coupling does not offer a suitable way to realize a high extinction ratio at a resonant wavelength. A theoretical study suggests that momentum space matching could be used to implement suitable coupling [19]. Up to now, the demonstrated side-coupled NPCC filters only show an extinction ratio (ER) with a transmission lower than 4.2 dB on resonance [20], which is insufficient for many applications such as switches, wavelength division demultiplexing (WMD) filters, modulators, and sensors. Using an S-bend bus waveguide is another way to implement efficient side-coupling of NPCC filters [15]. However, the insertion loss of the through wave will accumulate in the bus waveguide when multiple serially side-coupled NPCCs are involved.

This work demonstrates one-dimensional photonic crystal cavities on a ring (PCCoBW) on SOI. Utilizing the enhanced exposure of evanescent fields of bending resonant modes, we achieve filters with ERs around −40 dB and $Q > 10^4$. Furthermore, the coupling gap is relaxed to above 200 nm, which is more feasible for fabrication. As far as we know, these results represent the highest ER and Q observed in side-coupled NPCCs.

2. Proposal and Analysis

The schematic models of a side-coupled conventional and bending NPCC are shown in Figure 1a and Figure 1b, respectively. By using the temporal coupling mode theory, the transmission through the output port can be described as [21]:

$$T = \frac{(\omega - \omega_0)^2 + (\tau_r)^2}{(\omega - \omega_0)^2 + (\tau_c + \tau_r)^2} \tag{1}$$

where ω_0 is the resonant frequency of the side-coupled NPCC and τ_c and τ_r represent the coupling rates at which the NPCC couples to the bus waveguide and the radiation modes, respectively. The partial quality factor Q_i is related to the coefficients of the coupling rate by $\tau_i = \omega_0/2Q_i$, where i represents a certain coupling channel i and the total quality factor can be expressed as $Q_t = 1/(1/Q_c + 1/Q_r)$.

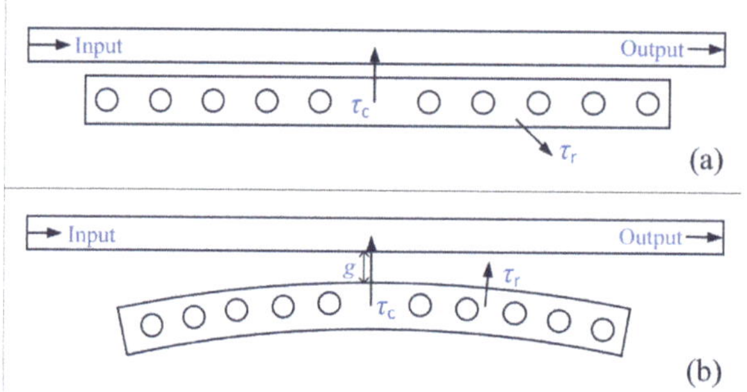

Figure 1. Schematics of the side-coupled (**a**) conventional NPCC and (**b**) bending NPCC.

Compared to the conventional side-coupled NPCC (Figure 1a), the proposed configuration attains a natural adiabatic coupling structure, which could efficiently decrease the reflection and the insertion loss of non-resonant waves. From Equation (1), to achieve complete extinction at resonance (critical coupling), one should ensure that $\tau_c \gg \tau_r$ or $Q_c \ll Q_r$. Our previous work [22] indicates that both the τ_c and τ_r of the structure in Figure 1a will increase when the coupling gap g decreases. Fundamentally, the bandgap of the compound periodic waveguide composed of the parallel bus waveguide and nanobeam will shrink and even disappear when g becomes small. It has been proven [20] that implementing a high-contrast-on-resonance notch filter using the parallel coupled configurations is intractable (Figure 1a). For critical coupling, sustaining a high Q_r for small gap g (small Q_c) is important. Figure 1b provides a possible design that alleviates the degradation of Q_r when g is small.

Firstly, we employ the FDTD simulation to investigate the influence of radius R of the bending nanobeam cavity on the partial Q_is of the bending NPCCs. In the simulations, we assume that the NPCCs are fabricated on a 220 nm silicon layer SOI platform. The bending nanobeam is 500 nm wide with a bending radius $R = N_R a$, where N_R is the normalized bending radius and a is the lattice constant. Here a is 340 nm, and the radii of the periodic holes vary linearly, starting from the hole with radius $r_1 = 0.317a$ and decreasing to both sides of the cavity to the radius of hole $r_N = 0.161a$. The calculation is started from a straight nanobeam photonic crystal cavity with a number of holes $N_h = 40$. The fundamental transverse electric (TE) polarization mode of this cavity shows an intrinsic $Q > 3 \times 10^6$. To investigate the influence on the Qs of the bending radius R and coupling gap g, we calculate the partial Q_i using the following equation

$$Q_i = (\omega_0 U)/P_i \tag{2}$$

where U is the energy of the resonant modes stored in the cavity, P_i is the power loss per cycle due to some of the coupling mechanisms in channel i, and ω_0 is the resonant frequency of the resonant mode. Figure 2 reveals the essential fact that Q of the unloaded bending NPCC could remain around the initial value even when the bending radius R decreases from 100,000a to 30a. The mode profiles of the conventional NPCC and bending NPCC with $N_R = 30$ are shown in Figure 3. The mode distribution (E intensity) of the NPCC with $N_R = 30$ does not show noticeable radiation loss compared to that of the straight nanobeam cavity.

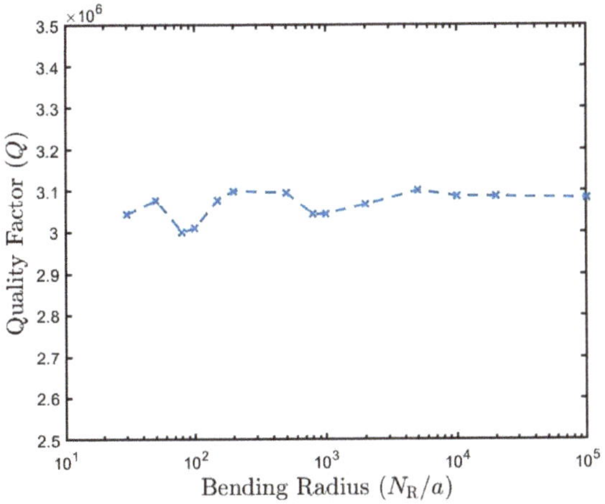

Figure 2. Influence of bending radius N_R on quality factor Q of the bending NPCs.

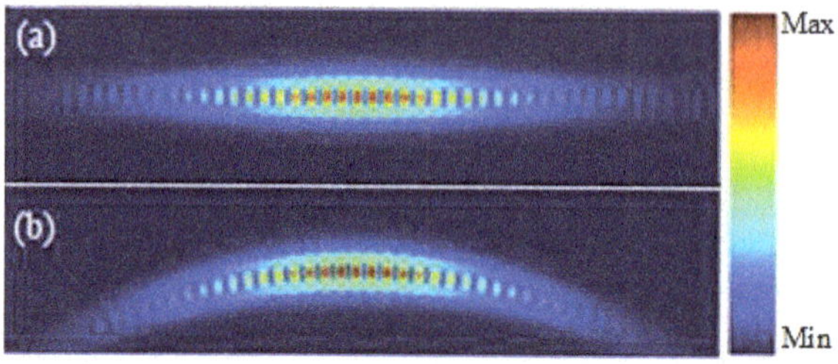

Figure 3. Mode profiles of (**a**) a conventional NPCC and (**b**) the bending NPCC.

To investigate the ER of the spectral dip at the resonance of the coupled bending NPCC, the calculated partial Qs according to Equation (2) are shown in Figure 4. At the coupling gap of 200 nm, Q_r increase gradually when the nanobeam cavity bends outside away from the bus waveguide, and achieve a maximum of 3×10^6 when N_R takes 1500. Further bending the nanobeam decreases the Q_r slightly since dramatic radiation loss occurs. However, Q_c sees a rapid increase when N_R changes from 50 to 4000 and gradually saturates to 2300 when N_R increases above 10,000. It is worth noting that a smaller bending radius $N_R < 30$ would increase the Q_c since an insufficient coupling region is ensured. Thus a moderately small N_R is critical for implementing high-efficient side-coupling. From the data in Figure 4b, choosing a bending radius between $30a$–$100a$ for ER > 30 dB is suitable.

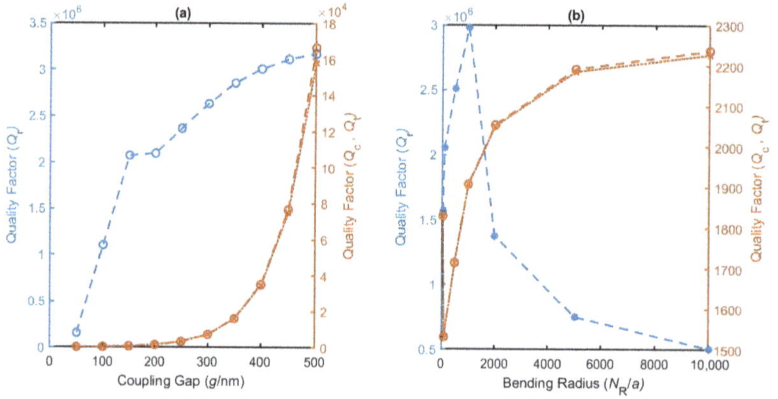

Figure 4. Partial Qs (Q_r: blue circled dash; Q_c: brown circled dash; Q_t: dotted) of the side-coupled bending NPCC as a function of (**a**) coupling gap g (with fixed $N_R = 100$) and (**b**) bending radius N_R (with fixed $g = 200$ nm).

When the N_R is fixed at 100, the dependencies of Q_c and Q_r on g are shown in Figure 4a. Both Q_c and Q_r rise with the increasing coupling gap g. Specifically, Q_r increases rapidly for smaller g until 200 nm, whereas Q_c increases rapidly when g is greater than 300 nm. In situations where $N_h/N_R = \pi$ (e.g., number of holes $N_h = 50$ and $N_R < 16$), the two ends of a bending nanobeam cavity overlap and naturally form a photonic crystal cavity on the bending waveguide (PCCoBW). It should be noted that the mechanism of the proposed PCCoBW is totally different from those photonic crystal ring resonators (PCRRs) in refs. [23–26], where a bending periodic (rotational symmetry) holey waveguide is used to replace the traditional waveguide of a ring resonator. By creating two cavity centers, the PCCoBW can simultaneously support dielectric and air resonant modes. In this case, the dielectric and air mode cavities share the same multiple periodic air holes, where the interplay between the two distinct resonances could increase in case of sufficient mode overlap.

3. Device Design

The proposed PCCoBWs on a 220 nm silicon layer are modeled and analyzed using the 3D FDTD with the waveguide parameters given in Section 2. N_h takes 50 and 70, corresponding to N_R of 15.91 and 22.28, respectively. The radii of air holes vary from $r_1 = 0.317a$ to $r_N = 0.161a$ and the lattice is 340 nm in both cases. The final design includes air and dielectric resonant modes coexisting in bending nanobeam cavities coupled, respectively, with two access waveguides. The parameters used in the simulation and the results are summarized and listed in Tables 1 and 2.

Table 1. Parameters used and results achieved in the simulations (N_h = 50, N_R = 15.92).

Mode Type	Order	Symmetry	Q	λ (μm)	$V[(\lambda/n)^3]$
Dielectric modes	D0	sym.	1.1×10^6	1.568	1
	D1	asym.	1.2×10^5	1.5592	1.75
	D2	sym.	6.3×10^4	1.5695	2.17
	D3	asym.	5.2×10^4	1.5786	2.61
Air modes	A0	asym.	7×10^6	1.4652	0.91
	A1	sym.	4.4×10^6	1.4337	1.63
	A2	asym.	1.3×10^6	1.4219	1.67
	A3	sym.	1.5×10^5	1.4016	2.03

Table 2. Parameters used and results achieved in the simulations (N_h = 70, N_R = 22.28).

Mode Type	Order	Symmetry	Q	λ (μm)	$V[(\lambda/n)^3]$
Dielectric modes	D0	sym.	2.4×10^6	1.5439	1.12
	D1	asym.	1.3×10^6	1.5533	1.9
	D2	sym.	5.3×10^5	1.5612	2.28
	D3	asym.	1×10^5	1.5693	2.69
Air modes	A0	asym.	8×10^6	1.4685	1.1
	A1	sym.	5.1×10^6	1.4465	1.7
	A2	asym.	2.3×10^6	1.4336	2.07
	A3	sym.	5.4×10^5	1.42	2.32

In the simulations, only the first four orders of TE polarization resonant modes are investigated, but they may represent the key metrics of these proposed PCCoBWs. Tables 1 and 2 show that modal volumes V of the designed cavities are comparable to those of conventional nanobeam photonic crystal cavities, typically with several times the $(\lambda/n)^3$, where λ is the resonant wavelength and n is the refractive index of the cavity center where the strongest electric field is located. The air mode cavity has a slightly lower V and higher Q than the dielectric modes since smaller holes are created in the cavity center, and better mode confinement is achieved. Both the air mode and dielectric mode have much slower V than those of photonic crystal ring resonators [23–26]. The mode profiles simulated by 3D FDTD of the dielectric and air modes (N_h = 50) are illustrated in Figure 5 and Figure 6, respectively. The corresponding mode numbers are also marked with the modes for comparison. These figures show that air and dielectric modes with high Q values are well confined in the bending PCCoBWs. In the profiles of high-order modes [Figures 5c,d and 6c,d], one can also observe the mode conversion from fundamental modes to first-order modes arising along the ring due to the waveguide bend. Comparing the mode profiles of air and dielectric modes, we can also see significant differences in mode-extending abilities along the ring. Specifically, the mode tail of the dielectric modes extends further to the opposite side of the ring than that of air modes.

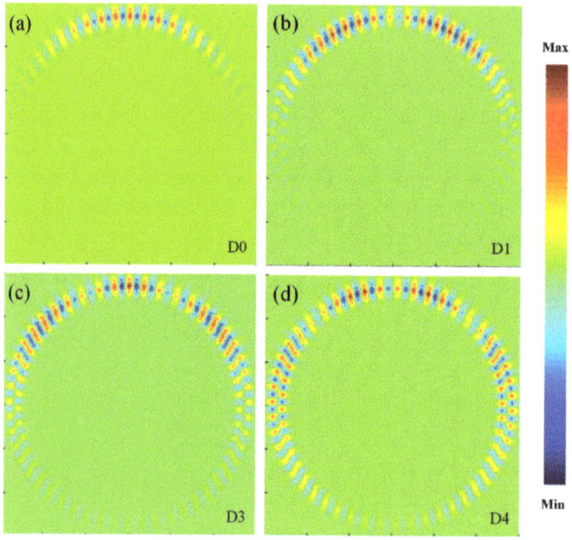

Figure 5. Mode profiles of the (**a**) D0, (**b**) D1, (**c**) D2 and (**d**) D3 order dielectric modes of the unloaded PCCoBW (N_h = 50).

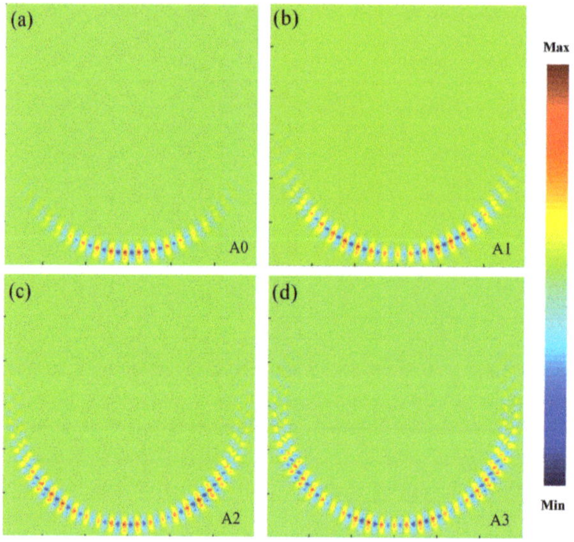

Figure 6. Mode profiles of the (**a**) A0, (**b**) A1, (**c**) A2 and (**d**) A3 order air modes of the unloaded PCCoBW (N_h = 50).

For the convenience of device characterization, two access waveguides with focusing sub-wavelength grating couplers are placed alongside the two cavity centers for coupling signal light into and out of the PCCoBW. To achieve better mode coupling, the access waveguides with various widths are chosen to match the effective refractive indexes of the air mode and dielectric mode of the holey waveguides, respectively. The details of the mode-matching design can be found in our earlier work [27].

4. Experiment and Discussion

A 100 keV electron beam lithography system and anisotropic plasma etch process are employed to provide smooth sidewalls and reduce the propagation loss of waveguides. Checking with a scanning electronic microscope (SEM) is conducted before a 2.2 μm thick cladding oxide layer is deposited on the top of etched devices. A typical SEM image of the fabricated PCCoBW (N_h = 50) is shown in Figure 7. Note that imperfection of the process would have influenced the critical size of the final devices, particularly the smaller holes. We also fabricated PCCoBWs with a larger bending radius (N_R = 22.28) and number (N_h = 70) of mirror air holes. The final silica-clad, fully etched 220 nm SOI devices were tested using a low-source spontaneous emission (SSE) tunable sweeping laser (Keysight 81960A, Santa Rosa, CA, USA) and a multi-channel optical power meter (Keysight N7745A, Santa Rosa, CA, USA). The response spectra of the through and drop ports which were excited from various input ports are plotted in Figures 8–10. The transmission spectrum of a straight waveguide has been used to calibrate the baseline of the response spectra.

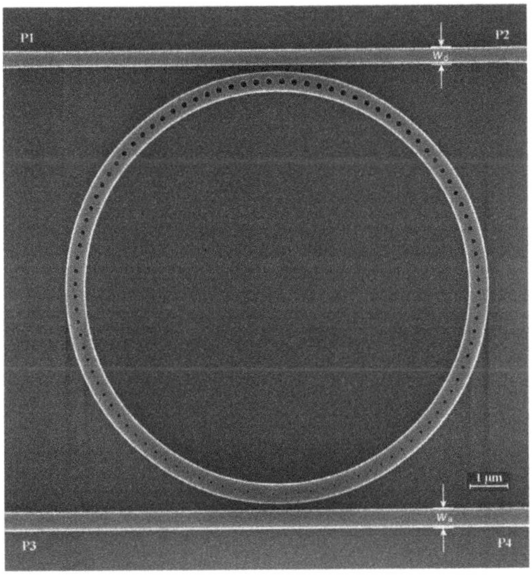

Figure 7. An SEM image of the fabricated PCCoBW on SOI (N_h = 50, a = 340 nm, w_d = 390 nm, w_a = 480 nm).

As shown in Figure 8, the fundamental air resonance presents a greater redshift ($\delta\lambda < 36$ nm) in spectra relative to the simulation results compared with those of the dielectric modes ($\delta\lambda < 7$ nm). It is also observed that as N_h increases (Figure 9), the measured resonance shift ($\delta\lambda$) in the air mode rises ($\delta\lambda > 80$ nm), while the wavelength shift in the dielectric mode remains essentially stable ($\delta\lambda \sim 16$ nm). The influences of fabrication errors on Q and λ_{res} have been systematically studied by Yamaguchi, Yuki et al. [28], and experimentally. Such substantial discrepancy has been confirmed to be mostly caused by the fabrication error of tiny structures such as small holes [15]. The layout for the fabricated device is a ring waveguide with an array of subtracted holes, which are defined in the Klayout design file as polygonal shapes. Each hole is constructed as a polygon with more than 150 vertices to achieve a high degree of geometric precision in the mask design. Therefore, the air mode is more susceptible to variations in the size of small holes that form the center of the air mode compared to the dielectric mode.

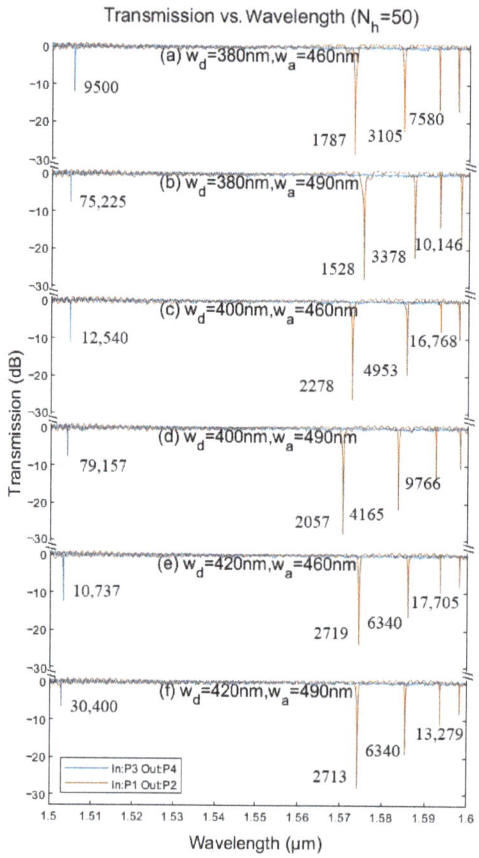

Figure 8. Measured spectra of the through ports of the fabricated PCCoBW with N_h = 50. The value labeled adjacent to the transmission dip corresponds to the Q factor of the resonance. (**a**) w_d = 380 nm and w_a = 460 nm, (**b**) w_d = 380 nm and w_a = 490 nm, (**c**) w_d = 400 nm and w_a = 460 nm, (**d**) w_d = 400 nm and w_a = 490 nm, (**e**) w_d = 420 nm and w_a = 460 nm, (**f**) w_d = 420 nm and w_a = 490 nm.

For the N_h = 50 case, an ER of more than 30 dB of the dielectric mode resonance is observed in the measured spectra, whereas a maximum ER of only 12 dB is found for air mode resonances. Comparing various subplots in Figure 8, one may find that the mode matching between the PCCoBW and the coupling waveguides plays significant roles in waveguide–cavity coupling coefficients, which substantially influence the ER of the corresponding resonance spectra. Specifically, for the dielectric mode coupling waveguide, the width w_d = 380 nm configuration provides the best coupling, while the waveguide width w_a = 460 nm is the most compelling case for air resonances. Similar results are also found in the measured spectra of the N_h = 70 case, as shown in Figure 9. The Q factors of these modes are also marked near the dip of the resonant spectra. The typical Q is 10^4 and 10^3 for the fundamental air and dielectric resonant modes, respectively. Since these Qs are measured from the coupled PCCoBWs, the intrinsic (standalone PCCoBW) Qs are supposed to be higher than 10^5.

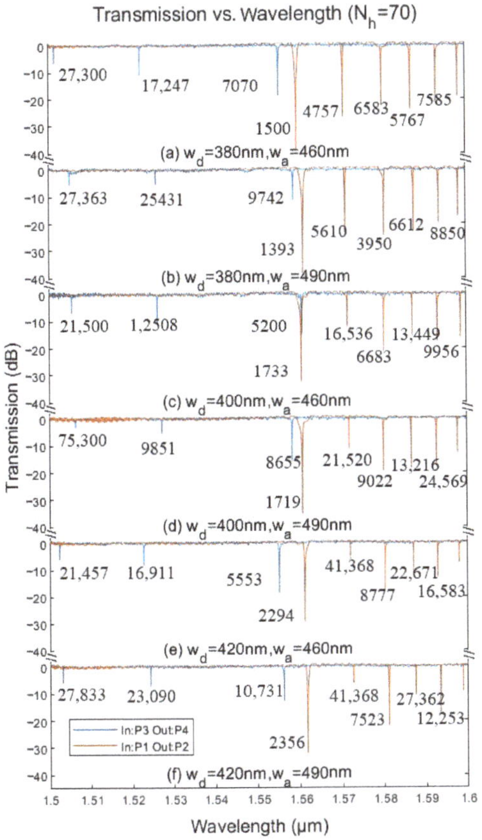

Figure 9. Measured spectra of the through ports of the fabricated PCCoBW with $N_h = 70$. The value labeled adjacent to the transmission dip corresponds to the Q factor of the resonance. (**a**) $w_d = 380$ nm and $w_a = 460$ nm, (**b**) $w_d = 380$ nm and $w_a = 490$ nm, (**c**) $w_d = 400$ nm and $w_a = 460$ nm, (**d**) $w_d = 400$ nm and $w_a = 490$ nm, (**e**) $w_d = 420$ nm and $w_a = 460$ nm, (**f**) $w_d = 420$ nm and $w_a = 490$ nm.

Figure 9a reveals an ER over 40 dB for the dielectric resonant mode at 1572.3 nm, representing the recorded ER observed in side-coupled PCCoBWs. The ER of the corresponding air mode resonances also rises to 20 dB. These improvements should be attributed to a larger bending radius, thus ensuring stronger coupling. Figure 9c–f also show another essential characteristic of the side-coupled bending PCCoBWs: mode-related coupling strength fluctuations. Combining the mode profiles in Figures 5 and 6 and noticing the differences between the symmetric and anti-symmetric modes, the resonance dips also reveal the corresponding coupling strengths and ERs. The distinct resonance dip indicates that mode overlapping in space is an essential factor influencing the coupling. When the mode is symmetric, the max field is located at the center of the coupling region and the max ER is obtained (Figure 9e,f), and vice versa.

We also measured the responses of the drop ports of these PCCoBWs, and the results are reported in Figure 10. It can be seen that mode-dependent coupling also plays a significant role in the responses of the drop ports. As shown in Figure 10a,d, no apparent air mode resonances can be found on the spectra of the drop ports when P3 serves as the input port, whereas high-order dielectric resonant modes are revealed when either P1 or P3 are used as the input port (Figure 10b,c). The reason causing this difference can be understood by comparing the mode profile distributions in Figures 5 and 6 for dielectric modes and air modes, respectively. The high-order dielectric modes extend to the cavity center of the other coupling region much more than the air modes, due to larger holes that lower confinement. The mode extension increases the mode overlay, resulting in a higher coupling strength with the drop waveguide.

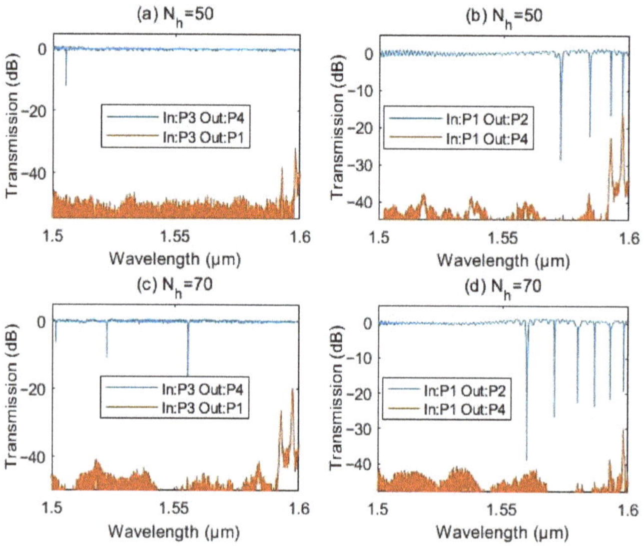

Figure 10. Measured spectra of the through and drop ports of the fabricated PCCoBWs with (**a**,**b**) N_h = 50 and (**c**,**d**) N_h = 70 holes.

5. Conclusions

In summary, 1D photonic crystal cavities on rings (PCCoBW) have been proposed and investigated. The air and dielectric modes of the bending PCCoBW can sustain high-quality factors approaching 10^5, even when the bending radius is less than $30a$. The experimental results conducted on SOI indicate that high ER over 40 dB and 20 dB can be achieved in the N_R = 22 case for dielectric and air resonant modes, respectively, when suitable mode matching is satisfied in dual-bus coupled PCCoBW configurations. The PCCoBW also shows the essential characteristics of mode-dependent coupling strength, distinct from conventional microring resonators or photonic crystal ring resonators. We hope that these findings will contribute to the development of low-power-consumption switches and high-performance filters and sensors, as well as engineering the interactions between air and dielectric resonant modes.

Author Contributions: Validation, H.Q. and D.K.; Investigation, P.Y.; Data curation, Z.W. and J.W.; Writing—original draft, P.Y., J.W. and F.G.; Writing—review & editing, F.G., H.Q. and D.K.; Supervision, P.C.; Project administration, P.C.; Funding acquisition, P.C. All authors have read and agreed to the published version of the manuscript.

Funding: This work was supported by the Scientific Research Program of the Zhejiang Provincial Department of Education (Y202456638), the Science and Technology Innovation Yongjiang 2035 Key Research and Development Program (2025Z089), and the Natural Science Foundation of Fujian (2023J01350).

Data Availability Statement: The original contributions presented in this study are included in the article. Further inquiries can be directed to the corresponding author.

Conflicts of Interest: The authors declare no conflicts of interest.

References

1. Yablonovitch, E. Inhibited Spontaneous Emission in Solid-State Physics and Electronics. *Phys. Rev. Lett.* **1987**, *58*, 2059–2062. [CrossRef]
2. Fink, Y.; Winn, J.N.; Fan, S.; Chen, C.; Michel, J.; Joannopoulos, J.D.; Thomas, E.L. A Dielectric Omnidirectional Reflector. *Science* **1998**, *282*, 1679–1682. [CrossRef] [PubMed]
3. Wu, F.; She, Y.; Jiang, H.; Liu, G.; Chen, G.; Xiao, S. Angle-immune strong coupling between two defect modes in a defective photonic hypercrystal. *Opt. Lasers Eng.* **2025**, *186*, 108842. [CrossRef]
4. Wei, Q.; Bi, D.; Qi, X.; Ren, M.; Wu, F. Angle-independent topological interface states in one-dimensional photonic crystal heterostructures containing hyperbolic metamaterials. *Opt. Lett.* **2025**, *50*, 451. [CrossRef]
5. Foresi, J.S.; Villeneuve, P.R.; Ferrera, J.; Thoen, E.R.; Steinmeyer, G.; Fan, S.; Joannopoulos, J.D.; Kimerling, L.C.; Smith, H.I.; Ippen, E.P. Photonic-bandgap microcavities in optical waveguides. *Nature* **1997**, *390*, 143–145. [CrossRef]
6. Deotare, P.B.; McCutcheon, M.W.; Frank, I.W.; Khan, M.; Loncar, M. High quality factor photonic crystal nanobeam cavities. *Appl. Phys. Lett.* **2009**, *94*, 121106. [CrossRef]
7. Velha, P.; Picard, E.; Charvolin, T.; Hadji, E.; Rodier, J.; Lalanne, P.; Peyrade, D. Ultra-High Q/V Fabry-Perot microcavity on SOI substrate. *Opt. Express* **2007**, *15*, 16090. [CrossRef] [PubMed]
8. Dong, G.; Xiong, M.; Dimopoulos, E.; Sakanas, A.; Semenova, E.; Yvind, K.; Yu, Y.; Mørk, J. Experimental demonstration of a nanobeam Fano laser. *Opt. Express* **2024**, *32*, 5242. [CrossRef]
9. Liu, L.; Ye, M.; Yu, Z.; Xue, W. Notch microwave photonic filter with narrow bandwidth and ultra-high all-optical tuning efficiency based on a silicon nanobeam cavity. *J. Light. Technol.* **2023**, *41*, 5051–5058. [CrossRef]
10. Yao, D.; Jiang, Z.; Zhang, Y.; Xie, H.; Wang, T.; Wang, J.; Gan, X.; Han, G.; Liu, Y.; Hao, Y. Ultrahigh thermal-efficient all-optical silicon photonic crystal nanobeam cavity modulator with TPA-induced thermo-optic effect. *Opt. Lett.* **2023**, *48*, 2325. [CrossRef] [PubMed]
11. Zhong, H.; Li, J.; He, Y.; Zhang, R.; Wang, H.; Shen, J.; Zhang, Y.; Su, Y. Ultra-low-power consumption silicon electro-optic switch based on photonic crystal nanobeam cavity. *npj Nanophoton.* **2024**, *1*, 33. [CrossRef]
12. Wang, J.; Wang, C.; Han, Z.; Tian, H. On-chip trapping and sorting of nanoparticles using a single slotted photonic crystal nanobeam cavity. *Opt. Express* **2022**, *30*, 11192. [CrossRef] [PubMed]
13. Gao, Y.; Dong, P.; Shi, Y. Suspended slotted photonic crystal cavities for high-sensitivity refractive index sensing. *Opt. Express* **2020**, *28*, 12272. [CrossRef] [PubMed]
14. Quan, Q.; Deotare, P.B.; Loncar, M. Photonic crystal nanobeam cavity strongly coupled to the feeding waveguide. *Appl. Phys. Lett.* **2010**, *96*, 203102. [CrossRef]
15. Yu, P.; Qiu, H.; Yu, H.; Wu, F.; Wang, Z.; Jiang, X.; Yang, J. High-Q and high-order side-coupled air-mode nanobeam photonic crystal cavities in silicon. *IEEE Photon. Technol. Lett.* **2016**, *28*, 2121–2124. [CrossRef]
16. Pernice, W.H.P.; Xiong, C.; Schuck, C.; Tang, H.X. High-Q aluminum nitride photonic crystal nanobeam cavities. *Appl. Phys. Lett.* **2012**, *100*, 091105. [CrossRef]
17. Yu, P.; Qiu, C.; Hu, T.; Qiu, H.Y.; Ge, F.F.; Jiang, X.Q.; Yang, J.Y. High- Q Photonic Crystal Cavity in a Single-Mode Silicon Ridge Waveguide. *Chin. Phys. Lett.* **2013**, *30*, 104204. [CrossRef]
18. Yu, P.; Qiu, H.; Dai, T.; Cheng, R.; Lian, B.; Li, W.; Yu, H.; Yang, J. Ultracompact Channel Add-Drop Filter Based on Single Multimode Nanobeam Photonic Crystal Cavity. *J. Light. Technol.* **2021**, *39*, 162–166. [CrossRef]
19. Afzal, F.O.; Halimi, S.I.; Weiss, S.M. Efficient side-coupling to photonic crystal nanobeam cavities via state-space overlap. *J. Opt. Soc. Am. B* **2019**, *36*, 585. [CrossRef]
20. Halimi, S.I.; Hu, S.; Afzal, F.O.; Weiss, S.M. Realizing high transmission intensity in photonic crystal nanobeams using a side-coupling waveguide. *Opt. Lett.* **2018**, *43*, 4260. [CrossRef]
21. Haus, H.A. *Waves and Fields in Optoelectronics*; Solid State Physical Electronics Series; Prentice-Hall: Englewood Cliffs, NJ, USA, 1984; p. 402.
22. Yu, P.; Hu, T.; Qiu, C.; Shen, A.; Qiu, H.; Wang, F.; Jiang, X.; Wang, M.; Yang, J. Ultracompact, Reflection-Free and High-Efficiency Channel Drop Filters Based on Photonic Crystal Nanobeam Cavities. *Chin. Phys. Lett.* **2013**, *30*, 034210. [CrossRef]
23. Urbonas, D.; Balčytis, A.; Vaškevičius, K.; Gabalis, M.; Petruškevičius, R. Air and dielectric bands photonic crystal microringresonator for refractive index sensing. *Opt. Lett.* **2016**, *41*, 3655. [CrossRef]
24. Lee, J.Y.; Fauchet, P.M. Slow-light dispersion in periodically patterned silicon microring resonators. *Opt. Lett.* **2011**, *37*, 58. [CrossRef] [PubMed]
25. Goldring, D.; Levy, U. Highly dispersive micro-ring resonator based on one dimensional photonic crystal waveguide design and analysis. *Opt. Express* **2007**, *15*, 3156–3168. [CrossRef]

26. Lu, X.; McClung, A.; Srinivasan, K. High-Q slow light and its localization in a photonic crystal microring. *Nat. Photon.* **2022**, *16*, 66–71. [CrossRef]
27. Yu, P.; Qiu, H.; Cheng, R.; Chrostowski, L.; Yang, J. High-Q antisymmetric multimode nanobeam photonic crystal cavities in silicon waveguides. *Opt. Express* **2018**, *26*, 26196. [CrossRef]
28. Yamaguchi, Y.; Jeon, S.W.; Song, B.S.; Tanaka, Y.; Asano, T.; Noda, S. Analysis of Q-factors of structural imperfections in triangular cross-section nanobeam photonic crystal cavities. *J. Opt. Soc. Am. B* **2015**, *32*, 1792. [CrossRef]

photonics

MDPI

Article

Silicon-Based Multimode Complex Bragg Gratings for Spectra-Tailored Filter

Xiuqiu Shen, Huifang Kang, Wangping Wang, Xiong Liang * and Huiye Qiu *

School of Physics and Mechanical and Electronical Engineering, Longyan University, Longyan 364012, China
* Correspondence: 82001008@lyun.edu.cn (X.L.); qiuhuiye@lyun.edu.cn (H.Q.)

Abstract

Multimode waveguide Bragg gratings (MWBGs) provide significant advantages over traditional single-mode counterparts through their mode-coupling operations. Nevertheless, flexible spectral response design methodologies for MWBG-based filter remain less studied. This work introduces a spectral tailoring methodology enabling physically realizable complex responses in MWBGs. We demonstrate silicon-based multi-channel Gaussian-shaped MWBGs using lateral phase delay modulation (LPDM) apodization. Experimental results confirm close conformance between measured spectral responses and target design specifications.

Keywords: integrated optics device; wavelength filtering devices; bragg reflectors

1. Introduction

Silicon Bragg gratings have become an essential component of integrated photonic devices, finding applications in optical filters, sensors, WDM networks, optical signal processing, dispersion compensation, microwave photonics, and so on [1–8]. To date, most of the research has focused on traditional Bragg gratings fabricated on single-mode waveguides. As a two-port device, Bragg gratings typically require an additional 3 dB optical coupler or a circulator to obtain the reflected light. Nevertheless, using a 3 dB coupler can introduce an additional 6 dB loss at the drop port as the reflected signal must pass through the coupler twice, and a circulator is difficult to be integrated on SOI platform due to the lack of CMOS-compatible magneto-optic materials.

Multimode waveguide Bragg gratings (MWBGs) have recently attracted significant attention, as they can address the limitations of traditional Bragg gratings and offer additional advantages [9,10]. A MWBG with a specifically designed grating period can couple the forward TE0 mode to a backward higher-order mode at the target wavelength. The reflected higher-order mode is then converted to the fundamental mode and efficiently dropped via an adiabatic coupler. Therefore, the need for an extra 3 dB optical coupler or circulators is eliminated. A MWBG and a pair of adiabatic couplers form a four-port device, thereby functioning as an add-drop filter. However, previous studies on MWBGs have primarily focused on controlling specific spectral characteristics, such as increasing the sidelobe suppression ratio, sharpening the edges of the reflection response, and narrowing the operational bandwidth [11–13]. In these efforts, the grating strength was modulated along the grating by adjusting the depths or positions of the sidewall grating corrugations. This technique is well known as apodization. However, exclusive modulation of the grating length is insufficient for realizing complex target spectra.

The synthesis of Bragg gratings has been first studied in the field of fiber optics. Several kinds of mathematical models and algorithms have been proposed for finding the grating

 https://doi.org/10.3390/photonics12090924

structure (the grating strength and phase) from a specified, complex spectrum for fiber Bragg grating [14–16]. Recently, complex spectral tailoring for sidelobe-suppressed filters, multi-channel filters and microwave photonics Hilbert transformers has been realized on the traditional Bragg gratings and Grating-assisted contra-directional couplers [17–19]. However, the spectral tailoring of MWBGs has not been proposed and demonstrated on a silicon-on-insulator (SOI) platform.

In this work, we present a methodology aimed at engineering the complex spectral response of silicon MWBGs. We demonstrate the efficacy of the methodology through the experimental validation of multi-channel Gaussian spectral shaping. This study enables flexible manipulation of spectral characteristics in the integrated silicon MWBG-based devices, positioning them as promising candidates for reconfigurable photonic components including couplers, wavelength-selective switches, and add-drop multiplexers with customized spectral functionalities [8,20–22].

2. Design Methodology

This paper presents a fundamental design methodology for realizing an arbitrary complex spectral response using a silicon-based MWBG, in which implementing a multi-channel Gaussian-shaped filter serves as an example. The design process begins with the determination of a target complex spectral response. However, the initial target response is often not physically realizable and must be modified to become physically realizable before proceeding with the subsequent grating design. Once the physically realizable response is achieved, the discrete layer-peeling algorithm (DLPA) can be utilized to accurately calculate the grating strength and phase profiles required to meet the target response. In the next step, several fundamental parameters of the MWBG (including waveguide width, grating corrugation width, corrugation shape and grating period) are defined. Subsequently, the grating structure is established by modulating both the grating strength and phase along its length, in accordance with their respective calculated target profiles. Once the complete grating structure is created, its spectral response should be simulated to verify consistency with the realizable target response. Finally, the designed gratings are fabricated and experimentally characterized. The measured results are compared with the realizable target spectrum, enabling the iterative refinement of the design process.

2.1. Formation the Target Spectrum

Typically, the ideal objective is to achieve a specific reflection spectral response. For a multi-channel dispersionless Gaussian spectral MWBG, the target spectrum $r(\lambda)$ can be expressed as

$$r(\lambda) = \sum_{i=1}^{n} r_i(\lambda) = A_1 \exp[-4\ln 2(\frac{\lambda - \lambda_{c1}}{\Delta\lambda_1})^2] + A_2 \exp[-4\ln 2(\frac{\lambda - \lambda_{c2}}{\Delta\lambda_2})^2] \quad (1)$$

where $r_i(\lambda)$ represents the reflectance of the i-th Gaussian peak at wavelength λ, A_i represents the peak reflectance, λ_c is the central wavelength, and $\Delta\lambda$ denotes the 3 dB bandwidth. It should be noted that a larger number of channels is enabled by setting n as $\geq$3.

Figure 1 illustrates the target multi-channel MWBG spectrum with Gaussian profiles, showing a channel separation of 4 nm. The peak wavelengths are 1555 nm and 1559 nm, respectively. Each channel's excess loss is 1.8 dB at the peak wavelength, and the 3 dB bandwidth of each channel is 1 nm.

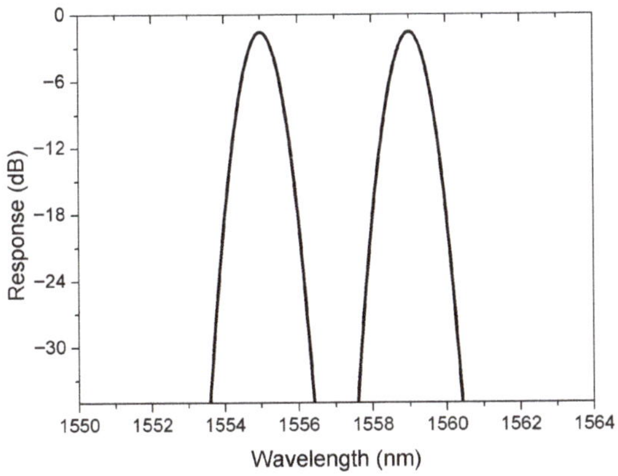

Figure 1. The target spectrum of the multi-channel MWBG.

As shown in Figure 2a, the original impulse response of the target spectrum is infinite in length and non-causal (i.e., it extends into the negative spatial domain). To address this, the target spectral response must be converted into a physically realizable form. We modify it by first truncating the original impulse response and then shifting the truncated response into the positive spatial domain. The resulting modified impulse response, shown in Figure 2b, is finite in length and causal.

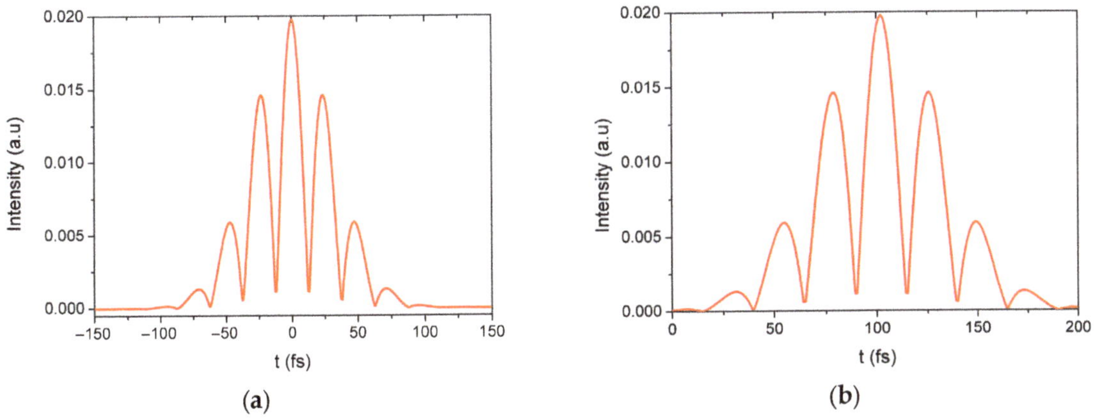

Figure 2. (**a**) The original impulse response, (**b**) the truncated and time-shifted impulse response of the target spectrum.

The physically realizable target response is obtained by applying the inverse Fourier transform to the modified impulse response. As shown in Figure 3, the realizable spectral response with a truncation window size of 100 fs (range 50 fs to 150 fs in Figure 2b) exhibits apparent degradation. Although the peak wavelengths match the designed spectral response, each channel suffers from 3.2 dB of excess loss, 1.2 nm of 3 dB bandwidth, and prominent sidelobes. This degradation is primarily caused by the insufficient truncation window size, which directly results in inadequate grating length. When the truncation window size increases to 200 fs (Figure 2b), the realizable spectral response closely matches the designed spectrum.

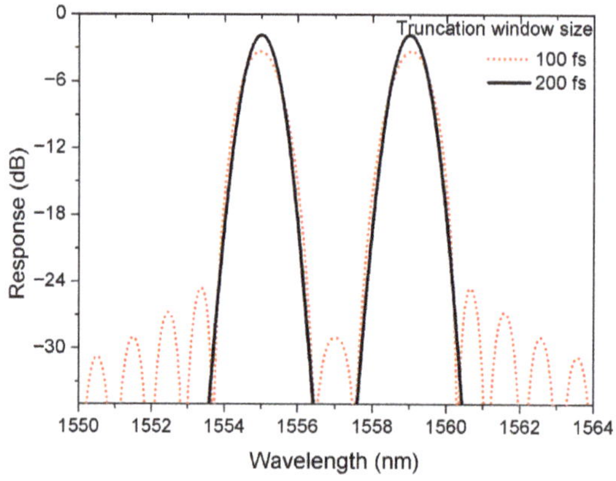

Figure 3. Physically realizable spectral responses of the designed multi-channel MWBG with truncation window size of 100 fs and 200 fs, respectively.

2.2. Calculating the Complex Coupling Coefficient

The DLPA is a computationally efficient inverse scattering method that has been widely adopted for fiber Bragg grating synthesis. Its principle relies on solving the coupled-mode equations using piecewise scattering and a propagation matrix approach. When the piecewise section length is sufficiently small, the grating's leading section can be determined solely by the impulse response's leading edge. Thus, the grating structure can be reconstructed sequentially layer by layer, while the complex coupling coefficient (including the grating strength $\kappa(z)$ and phase $\varphi(z)$ distribution along the grating) is directly obtained via the DLPA method. Figure 4 shows the grating strength and phase distribution synthesized derived from the realizable target spectral response. The $\kappa(z)$ profile exhibits significantly greater complexity than conventional MWBGs. Meanwhile, the required $\varphi(z)$ distribution manifests discrete π-phase shifts, corresponding to a dispersionless MWBG configuration [18]. These structural characteristics necessitate a high-performance and reliable apodization technique for reliable grating realization.

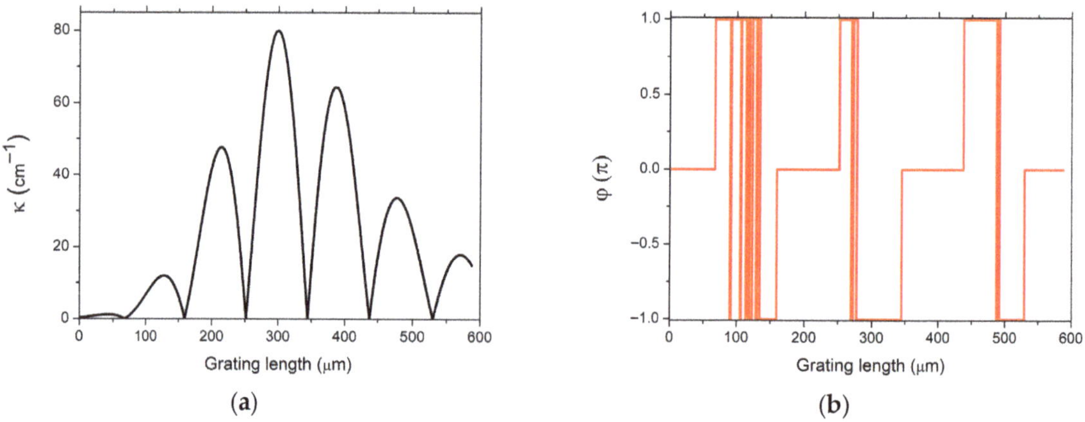

Figure 4. (**a**) Grating strength $\kappa(z)$ and (**b**) phase $\varphi(z)$ distributions along the grating calculated by using the DLPA.

2.3. Structure Design

The structure of the complex-synthesized MWBG proposed is schematically illustrated in Figure 5. The 220 nm-thick silicon waveguide is embedded in a 2-μm buried oxide layer with a 2.2-μm silicon dioxide upper cladding. The multimode waveguide width W in the MWBG is configured at 850 nm to enable simultaneous propagation of both TE0 and TE1 modes. This grating structure employs rectangular sidewall modulations along the waveguide to achieve contra-directional coupling from the TE0 mode to the TE1 mode at selected wavelengths. For the target peak wavelength, the grating period Λ is determined as 314 nm via the phase matching condition.

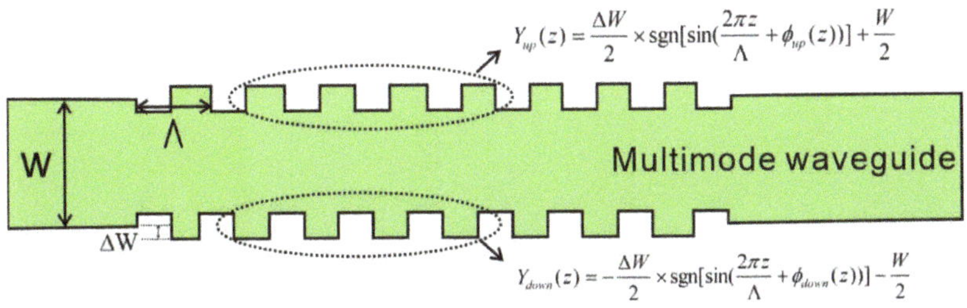

Figure 5. Structure of a complex-synthesized MWBG.

The spectral response engineering hinges on the precise spatial modulation of both κ and φ of the grating, as defined by their respective distributions $\kappa(z)$ and $\varphi(z)$. A uniform MWBG can be treated as a superposition of two identical sub-gratings formed on either side of the multimode waveguide with a fixed phase offset of $\Delta\phi$. To achieve precise κ modulation, the lateral phase delay modulation (LPDM) method is implemented. This establishes the relationship between κ and $\Delta\phi$ through the equation:

$$\kappa(\Delta\phi) = \kappa_{\max}|\sin(\Delta\phi/2)| \tag{2}$$

where $\kappa_{\max}$ is the maximum coupling efficient when $\Delta\phi$ is π. The phase shift distribution $\Delta\phi(z)$ is then determined by the equation:

$$\Delta\phi(z) = 2\sin^{-1}(\kappa_n(z)) \tag{3}$$

where $\kappa_n(z)$ is the normalized $\kappa(z)$.

The phase components of the spatial functions of the top and bottom sub-gratings are formed by the superposition of the phase delay term and $\varphi(z)$:

$$\phi_{up}(z) = \Delta\phi(z)/2 + \varphi(z) \tag{4}$$

$$\phi_{down}(z) = -\Delta\phi(z)/2 + \varphi(z) \tag{5}$$

Therefore, the spatial edge functions of the two sides of the MWBG can be expressed as follows:

$$Y_{up}(z) = \frac{\Delta W}{2} \times \text{sgn}[\sin(\frac{2\pi z}{\Lambda} + \phi_{up}(z))] + \frac{W}{2} \tag{6}$$

$$Y_{down}(z) = -\frac{\Delta W}{2} \times \text{sgn}[\sin(\frac{2\pi z}{\Lambda} + \phi_{down}(z))] - \frac{W}{2} \tag{7}$$

where ΔW is the grating corrugation width.

The spectral response of the designed MWBG was calculated using a three-dimensional (3D) finite-difference time-domain (FDTD) tool [20]. The designed MWBG features a grating length of 587.76 μm (truncation window size: 200 fs) and a corrugation width of 10 nm, targeting a maximum coupling coefficient κ_{max} of 80 cm^{-1}. The calculated drop-port response, shown in Figure 6, exhibits good agreement with the realizable target spectral response presented in Figure 2.

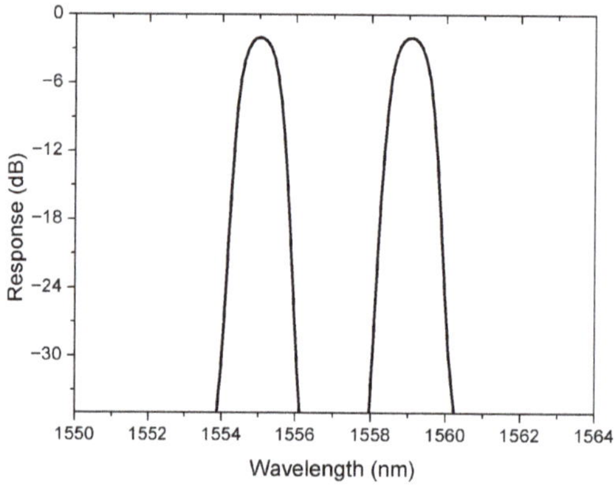

Figure 6. Three-dimensional FDTD calculated spectral response of the designed MWBG.

3. Experiment Result

The filter was fabricated on a commercial SOI wafer via electron-beam lithography (EBL) and plasma etching. Optical characterization employed a fiber probe station with a tunable laser source for signal injection and optical power sensors for output detection. Figure 7a schematically shows the characterization circuit for the designed MWBG-based filter. An adiabatic coupler was designed to convert the backward-propagating TE1 mode into the TE0 mode at the drop port. Figure 7b shows a scanning electron microscope (SEM) image of the fabricated MWBG.

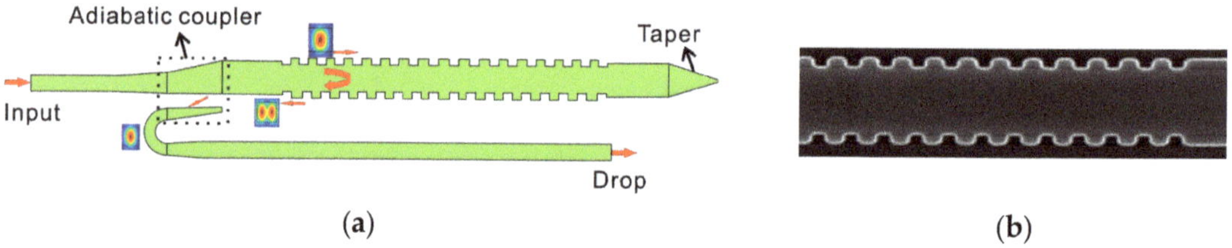

Figure 7. (**a**) Schematic representation of the testing circuit for a MWBG. (**b**) SEM image of the fabricated MWBG.

Figure 8 shows the measured normalized spectral response after calibrating out the insertion losses of the input and output grating couplers (GCs). The response exhibits a multi-channel Gaussian profile consistent with the target response with the designed 4 nm channel separation achieved. However, the measured peak wavelengths (1567 nm and 1571 nm) exhibit a redshift relative to design values, and the 3 dB bandwidth of each channel (1.3 nm) is slightly wider than intended. Based on the Bragg condition, the redshift primarily results from an increase in the effective refractive index. Our further analysis, supported by SEM measurements, indicates that the unintended enlargement of the waveguide width during fabrication is the dominant factor contributing to the redshift. The sidelobe suppression ratio of 18 dB and the excess loss of 1.4 dB are also measured.

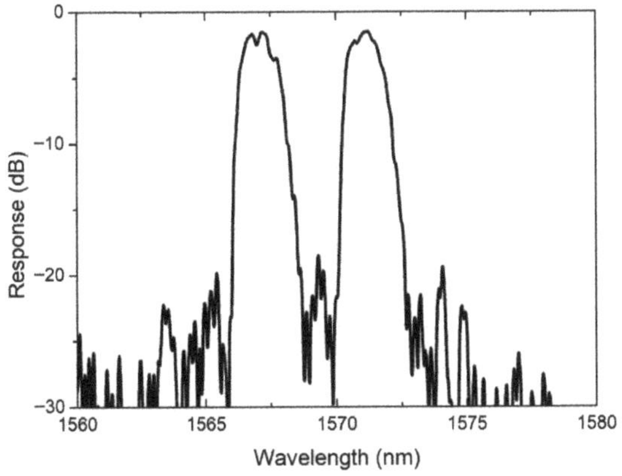

Figure 8. Measured normalized drop-port spectrum of the MWBG with $\Delta W = 10$ nm.

To investigate the influence of the grating strength $\kappa(z)$ and phase $\varphi(z)$, we also fabricated two MWBG variants: (1) grating maintaining $\kappa(z)$ while with $\varphi(z) = 0$ and (2) grating with enhanced $\kappa(z)$ while preserving $\varphi(z)$.

Figure 9 presents the measured normalized spectral response of a MWBG with identical $\kappa(z)$ to that in Figure 4a but with $\varphi(z) = 0$. The resulting spectrum exhibits seven channels showing non-uniform bandwidths and insertion losses, demonstrating significant differences from the target spectral response. This confirms that the simultaneous engineering of both grating strength and phase distributions is essential to achieve specific reflection spectral responses.

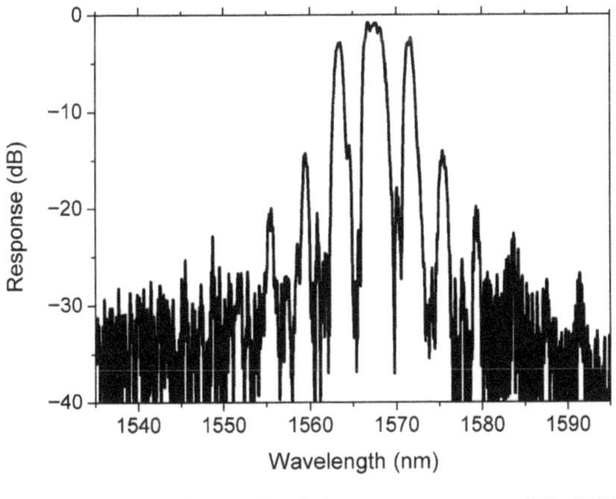

Figure 9. Measured normalized drop-port spectrum of the MWBG with $\Delta W = 10$ nm, $\varphi(z) = 0$.

To enhance $\kappa(z)$, the corrugation width was increased to 16 nm and 20 nm in two respective MWBGs, while maintaining the $\varphi(z)$ profile identical to that in Figure 4b. As shown in Figure 10, the measured normalized spectra for both devices exhibited the targeted multi-channel Gaussian profile. The measured 3 dB bandwidths are 2.6 nm and 2.0 nm, respectively, with channel separations of 4.1 nm. Although the stronger grating strength broadens the spectra compared with Figure 8, it does not alter the spectral profile shape. Consequently, adjusting ΔW provides a means to mitigate the influence of fabrication errors.

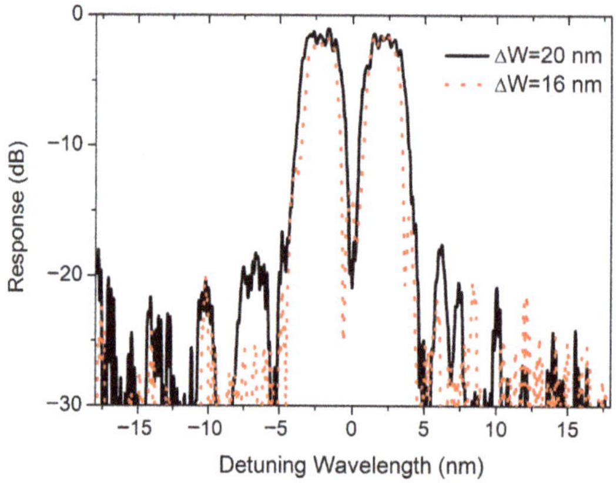

Figure 10. Measured normalized drop-port spectra of the MWBG with ΔW = 16 nm and ΔW = 20 nm, respectively.

4. Conclusions

In summary, we have presented a methodology for tailoring the complex responses of MWBGs. The comprehensive design flow is introduced, demonstrating its effectiveness through the realization of a multi-channel Gaussian-shaped SOI-based MWBG filter by implementing LPDM apodization. This approach enables flexible control over the complex responses of MWBG-based devices, such as couplers, switches, and add-drop multiplexers, allowing for the implementation of customized spectral functionalities.

Author Contributions: Conceptualization, X.S.; methodology, X.L.; investigation, H.K.; data curation, W.W.; writing—original draft preparation, X.S.; writing—review and editing, H.Q. All authors have read and agreed to the published version of the manuscript.

Funding: This research was funded by the Natural Science Foundations of Fujian Province (2023J01350, 2023J01968, 2023J01974, 2025J011705, 2025J011707) and the Xinluo District Industry Academia-Research Technology Joint Innovation Project (2022XLXYZ004).

Data Availability Statement: The original contributions presented in this study are included in the article. Further inquiries can be directed to the corresponding author.

Conflicts of Interest: The authors declare no conflicts of interest.

References

1. Murphy, T.E.; Hastings, J.T.; Smith, H.I. Fabrication and characterization of narrow-band Bragg-reflection filters in silicon-on-insulator ridge waveguides. *J. Light. Technol.* **2001**, *19*, 1938–1942. [CrossRef]
2. Wang, X.; Flueckiger, J.; Schmidt, S.; Grist, S.; Fard, S.T.; Kirk, J.; Doerfler, M.; Cheung, K.C.; Ratner, D.M.; Chrostowski, L. A silicon photonic biosensor using phase-shifted Bragg gratings in slot waveguide. *J. Biophotonics* **2013**, *6*, 821–828. [CrossRef] [PubMed]
3. Tan, D.T.H.; Ikeda, K.; Saperstein, R.E.; Slutsky, B.; Fainman, Y. Chip-scale dispersion engineering using chirped vertical gratings. *Opt. Lett.* **2008**, *33*, 3013–3015. [CrossRef] [PubMed]
4. Shi, W.; Yun, H.; Lin, C.; Greenberg, M.; Wang, X.; Wang, Y.; Fard, S.; Flueckiger, J.; Jaeger, N.; Chrostowski, L. Ultra-compact, flat-top demultiplexer using anti-reflection contra-directional couplers for CWDM networks on silicon. *Opt. Express* **2013**, *21*, 6733–6738. [CrossRef]
5. Simard, A.; Strain, M.; Meriggi, L.; Sorel, M.; LaRochelle, S. Bandpass integrated Bragg gratings in silicon-on-insulator with well-controlled amplitude and phase responses. *Opt. Lett.* **2015**, *40*, 736–739. [CrossRef]
6. Yao, J. Microwave Photonics. *IEEE J. Light. Technol.* **2009**, *27*, 314–335. [CrossRef]
7. Burla, M.; Cort'es, L.R.; Li, M.; Wang, X.; Chrostowski, L.; Azaña, J. Integrated waveguide Bragg gratings for microwave photonics signal processing. *Opt. Express* **2013**, *21*, 25120–25147. [CrossRef]

. Zhang, W.; Yao, J. A fully reconfigurable waveguide Bragg grating for programmable photonic signal processing. *Nat. Commun.* **2018**, *9*, 1396. [CrossRef] [PubMed]

. Qiu, H.; Yu, H.; Hu, T.; Jiang, G.; Shao, H.; Yu, P.; Yang, J.; Jiang, X. Silicon mode multi/demultiplexer based on multimode grating-assisted couplers. *Opt. Express* **2013**, *21*, 17904–17911. [CrossRef]

0. Qiu, H.; Jiang, J.; Yu, P.; Dai, T.; Yang, J.; Yu, H.; Jiang, X. Silicon band-rejection and band-pass filter based on asymmetric Bragg sidewall gratings in a multimode waveguide. *Opt. Lett.* **2016**, *41*, 2450–2453. [CrossRef]

1. Qiu, H.; Lin, L.; Yu, P.; Dai, T.; Jiang, X.; Yu, H. Narrow-Band Add-Drop Filter Based on Cladding-Modulated Apodized Multimode Bragg Grating. *J. Light. Technol.* **2019**, *37*, 5542–5547. [CrossRef]

2. Liang, X.; Cheng, R.; Shen, X.; Yu, P.; Dai, T.; Qiu, H. Spectral-distortionless, flat-top, drop-filter based on complementarily-misaligned multimode-waveguide Bragg gratings. *J. Light. Technol.* **2021**, *39*, 5896–5901. [CrossRef]

3. Ning, N.; Zhang, Q.; Wei, Z.; Wang, Q.; Xia, P.; Wang, Y. Ultra-Broadband Optical Filter Based on Step-Chirped Multimode Waveguide Bragg Grating. *IEEE Photon. J.* **2025**, *17*, 6601705. [CrossRef]

4. Feced, R.; Zervas, M.N.; Miguel, M. An efficient inverse scattering algorithm for the design of nonuniformfibre Bragg gratings. *IEEE J. Quantum Electron.* **1999**, *35*, 1105–1115. [CrossRef]

15. Skaar, J.; Wang, L.; Erdogan, T. On the synthesis of fiber gratings by layer-peeling. *IEEE J. Quantum Electron.* **2001**, *37*, 165–173. [CrossRef]

6. Bland-Hawthorn, J.; Buryak, A.; Kolossovski, K. Optimization algorithm for ultrabroadband multichannel aperiodic fiber Bragg grating filters. *J. Opt. Soc. Am. A* **2008**, *25*, 153–158. [CrossRef]

17. Cheng, R.; Chrostowski, L. Multichannel photonic Hilbert transformers based on complex modulated integrated Bragg gratings. *Opt. Lett.* **2018**, *43*, 1031–1034. [CrossRef]

18. Cheng, R.; Zheng, Y.; Feng, J.; Wang, L. Spectral tailoring of silicon grating-assisted contra-directional couplers. *Opt. Lett.* **2024**, *49*, 7210–7213. [CrossRef]

19. Xie, S.; Zhan, J.; Hu, Y.; Zhang, Y.; Veilleux, S.; Bland-Hawthorn, J.; Dagenais, M. Add–drop filter with complex waveguide Bragg grating and multimode interferometer operating on arbitrarily spaced channels. *Opt. Lett.* **2018**, *24*, 6045–6048. [CrossRef] [PubMed]

20. Fernández-Hinestrosa, A.; Luque-González, J.M.; Cheben, P.; Schmid, J.H.; Wang, S.; Wangüemert-Pérez, J.G.; Molina-Fernández, I.; Ortega-Moñux, A. Nanophotonic Bragg grating assisted Mach–Zehnder interferometers for O-band add-drop filters. *Sci. Rep.* **2024**, *14*, 18492. [CrossRef] [PubMed]

21. Praena, J.Á.; Carballar, A. Integrated Bragg Grating Spectra. *Photonics* **2025**, *12*, 351. [CrossRef]

22. Liu, D.; Wu, H.; Dai, D. Silicon Multimode Waveguide Grating Filter at 2 μm. *J. Light. Technol.* **2019**, *37*, 2217–2222. [CrossRef]

Article

Frequency Stabilization of Wideband-Tunable Low-Phase-Noise Optoelectronic Oscillator Based on Fundamental and Subharmonic RF Injection Locking

Zhihao Zhang [1,2,3], Dan Lu [1,2,3,*], Hao Song [1,2,3], Fei Guo [1,2,3] and Lingjuan Zhao [1,2,3]

1 State Key Laboratory of Optoelectronic Materials and Devices, Institute of Semiconductors, Chinese Academy of Sciences, Beijing 100083, China; zhangzhihao20@semi.ac.cn (Z.Z.); songhao@semi.ac.cn (H.S.); guofei11@semi.ac.cn (F.G.); ljzhao@semi.ac.cn (L.Z.)
2 Center of Materials Science and Optoelectronics Engineering, University of Chinese Academy of Sciences, Beijing 100049, China
3 Beijing Key Laboratory of Low Dimensional Semiconductor Materials and Devices, Beijing 100083, China
* Correspondence: ludan@semi.ac.cn

Abstract: A frequency stabilization scheme for a wideband-tunable optoelectronic oscillator (OEO) based on fundamental and subharmonic RF injection locking is proposed, achieving a tuning range of 2–22 GHz with low phase noise. The injection-locked performance of the OEO using the fundamental RF signal and its $1/n$ subharmonic is investigated. The fundamental injection locking achieves a phase noise of <−130 dBc/Hz @ 10 kHz offset across the entire tuning range. An examination of phase noise behavior at different subharmonic orders reveals that fundamental and subharmonic injection locking achieve a five-order-of-magnitude improvement in Allan variance (0.1 s) and approximately 40 dB phase noise reduction at a 10 Hz offset from the carrier. This approach leverages the low-phase-noise advantage of the OEO while benefiting from the high stability of low-frequency external RF sources, enabling multi-frequency point frequency stabilization optimization.

Keywords: optoelectronic oscillator; subharmonic injection locking; frequency stability; phase noise

1. Introduction

Optoelectronic oscillators (OEOs) can generate RF signals with high spectral purity and ultra-low phase noise [1] due to the use of high-quality (Q)-factor energy storage media. The phase noise of the RF signals generated by an OEO is independent of the oscillation frequency. This unique characteristic, combined with its exceptional features—including its high operational frequency, wideband tunability, low phase noise, and superior spectral purity—empower the OEO with unparalleled advantages in diverse applications in communication, sensing, ranging, and signal processing [2–5]. Nevertheless, to maintain low-phase-noise characteristics, an OEO typically needs optical fibers ranging from several kilometers to tens of kilometers. However, long optical fibers are highly susceptible to temperature and stress variations, which can cause mode drifting and hopping in the OEO's output and degrade its frequency stability. To effectively enhance OEOs' frequency stability, various technical approaches have been explored, including multi-loop OEO structures [6], high-Q RF photonic filters [7], phase-locked loops [8], parity–time symmetry [9], and injection-locking technology [10], among others. Among these, injection locking stands out as a simple yet effective approach, which usually involves injecting an RF signal into an OEO to lock the free-running oscillation frequency. It is possible to stabilize a high-performance OEO with a low-cost RF synthesizer. When the center frequency of the injected signal is tuned, the output of the OEO can be correspondingly adjusted, thereby achieving a frequency-tuned OEO. In addition to the frequency stability optimization mentioned above, the OEO system based on the injection-locking structure is also helpful for improving the side mode suppression ratio [11,12]. In such injection-locked OEO architectures, however, the microwave filter is usually excluded to allow for frequency tuning. In the absence of the filter, the gain and mode competition arising from the system's uneven response limit the tuning range, typically by one to several GHz.

In this paper, we propose a wideband-tunable injection-locked OEO structure that utilizes an Yttrium Iron Garnet (YIG) tunable filter and employs fundamental and subharmonic injection using an external tunable RF synthesizer. The OEO is tunable from 2 GHz to 22 GHz, whilst it maintaining a phase noise performance better than −130 dBc/Hz at a 10 kHz offset across the entire operational bandwidth. Injection locking serves as the primary mechanism for achieving frequency stabilization. To characterize the RF frequency stability, Allan variance is used. Injection locking significantly improves the OEO's Allan variance, demonstrating enhanced frequency stability. The Allan variance of the signal at 0.1 s has been improved by nearly five orders of magnitude, and the phase noise near the carrier frequency has been reduced by approximately 40 dB. Additionally, we investigate the effects of different subharmonic orders and the power of the injected subharmonic signal for 10 GHz RF generation, providing insights into how these factors influence the overall performance of the oscillator. The use of YIG helps maintain an optimal sidemode suppression ratio (SMSR), mode selectivity, and stability. Additionally, the current tuning mechanism of the filter supports fast frequency tuning, enhancing the system's overall responsiveness. This scheme enables a wideband-tunable OEO with superior frequency stability while preserving the low-cost and simplicity of the injection-locking approach.

2. Principle and Experimental Setup

For an injection-locked OEO, its phase noise power spectral density can be expressed as follows [10]:

$$S_{\mathrm{inj-oeo}}(\omega) = |H_{\mathrm{in}}(j\omega)|^2 S_{\mathrm{in}}(\omega) + |H_{\mathrm{osc}}(j\omega)|^2 S_{\mathrm{osc}}(\omega) \tag{1}$$

where $S_{\mathrm{in}}(\omega)$ and $S_{\mathrm{osc}}(\omega)$ represent the phase noise power spectral density of the injected signal and the free-running OEO, respectively. $|H_{\mathrm{in}}(j\omega)|^2$ and $|H_{\mathrm{osc}}(j\omega)|^2$ represent the phase noise transfer function of the injected RF signal in the injection-locked OEO and the phase noise transfer function of the free-running OEO in the injection-locked state, respectively, which can be expressed as

$$\begin{aligned} |H_{\mathrm{in}}(j\omega)|^2 &= \frac{S_{\mathrm{out}}(\omega)}{S_{\mathrm{in}}(\omega)} = \frac{1}{1+(\omega/\Delta\omega_0)^2} \\ |H_{\mathrm{osc}}(j\omega)|^2 &= \frac{S_{\mathrm{out}}(\omega)}{S_{\mathrm{osc}}(\omega)} = \frac{1}{1+(\Delta\omega_0/\omega)^2} \end{aligned} \tag{2}$$

where $\Delta\omega_0$ is the injection-locking range [13]:

$$\Delta\omega_0 = \frac{\omega_0}{2Q}\frac{U_{in}}{U_{out}} = \frac{f_{FSR}}{2}\sqrt{\frac{P_{in}}{P_{out}}} \tag{3}$$

Here, Q represents the quality factor, ω_0 represents the oscillation frequency, U_{in} and U_{out} represent the amplitudes of the input signal and the output signal, respectively, and P_{in} and P_{out} represent the powers of the input signal and the output signal, respectively, and f_{FSR} is the loop mode spacing. This shows that the locking range of the injection-locked OEO is determined by the loop mode spacing of the free-running OEO and the voltage ratio of the injected signal to the mode of the free-running OEO, where the loop mode spacing of the OEO is determined by the fiber length.

Figure 1 plots the relationship between $S_{inj-oeo}$ and ω using Equation (1). It can be seen that the phase noise of the injection-locked OEO system is determined by the collective contributions from both the free-running OEO and the injected signal. In the vicinity of the carrier frequency ($\omega << \Delta\omega_0$), the phase noise is predominantly influenced by the external injected RF signal. Conversely, in the frequency range far from the carrier ($\omega >> \Delta\omega_0$), the phase noise primarily reflects the characteristics of the free-running OEO. By using this approach, the low frequency stability follows that of the external RF source, while the phase noise maintains the OEO's inherent low-phase noise characteristics. In other words, the injection scheme combines the excellent frequency stability of conventional RF synthesizers with the superior phase noise performance offered by OEO technology.

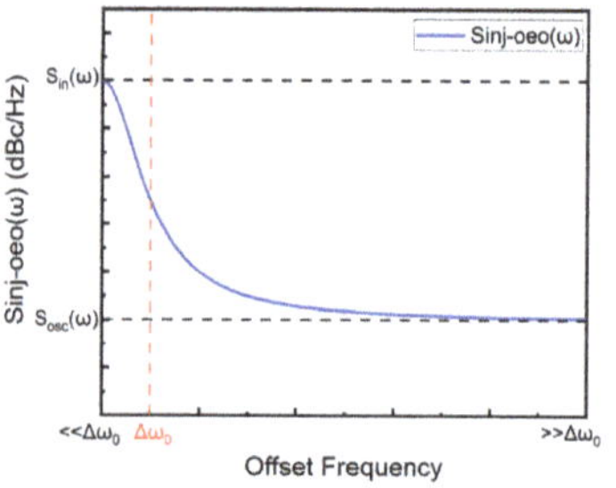

Figure 1. Effect of injection-locking range on phase noise of injection-locked OEO system.

Figure 2 shows the schematic configuration of the subharmonic RF injection-locked tunable OEO. The system employs a co-packaged module that integrates a DFB laser with a Mach–Zehnder modulator (−3 dB bandwidth of 20 GHz) for optical carrier generation and modulation purposes. The light from the DFB-MZM module is split into two parts by a 5:5 optical coupler (OC) and then converted into an RF signal using a balanced photodetector (BPD, Finisar BPDV2120R, Sunnyvale, CA, USA) after passing through optical fibers of 5 km and 300 m, respectively. The resulting RF signal is then amplified by two electrical amplifiers (EAs, Marki APM-6848PA, Greensboro, NC, USA) with a total gain of 46 dB. After amplification, the signal passes through a DC block and is filtered by a tunable YIG filter (tunable range: 1.8–23 GHz; bandpass width: approximately 55 MHz, tuning rate of 75 MHz/mA). Subsequently, the RF signal is split into two paths by a 9:1 electrical coupler (EC). Ninety percent of the signal serves as the output RF signal of the OEO, while the remaining 10% is further amplified by a third EA (gain: 23 dB) before being modulated onto the MZM in the DFB-MZM module. By adjusting the control current of the YIG filter, its center frequency can be tuned, thereby enabling the tunability of the OEO system's oscillation frequency. The optical output of the OEO is analyzed using an optical spectrum analyzer (Advantest Q8384, Tokyo, Japan, minimum resolution: 0.01 nm), while the phase noise and Allan variance are measured using a phase noise analyzer (Rohde & Schwarz FSWP50, Munich, Germany, frequency range: 1 MHz to 50 GHz).

Compared to conventional injection-locking structures, the proposed system incorporates a tunable filter to ensure effective injection locking, mitigating the impact of the system's uneven frequency response. This design helps maintain an optimal SMSR, mode selectivity, and stability. Additionally, the current tuning mechanism of the filter supports fast frequency tuning, enhancing the system's overall responsiveness. In this structure, the YIG filter is employed for coarse frequency selection, while the injected RF signal enables precise frequency tuning. The RF source, operating at the $1/n$ subharmonic of the OEO's free-running frequency, allows the OEO to lock onto the n-th harmonic of the injected RF signal, which is generated through the nonlinear frequency response of the optical modulator.

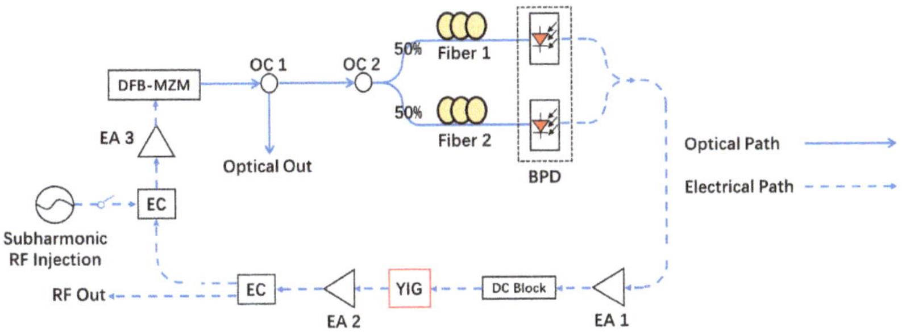

Figure 2. Schematic of wideband-tunable OEO based on subharmonic RF injection locking. (OC: optical coupler; BPD: balanced photodetector; EA: electrical amplifier; YIG: Yttrium Iron Garnet Filter; EC: electrical coupler.)

3. Experiments and Results

3.1. Fundamental RF Signal Injection Locking

The baseline performance of the tunable OEO prior to injection locking was first evaluated. Figure 3a,b show the RF spectrum for different spans and Figure 3c shows the phase noise of the 10 GHz output. The RF signal exhibits high spectral purity with a spurious suppression ratio of approximately 75.9 dB. The side mode spacing of $\Delta\nu = 660$ kHz corresponds to a short loop length of 300 m. The phase noise reaches −133.83 dBc/Hz at a 10 kHz offset from the carrier frequency. By adjusting the current of the YIG filter, the free-running OEO can be tuned within the range of 2–22 GHz. Figure 4a presents the RF spectrum of the tunable OEO with a 1 GHz frequency step, while Figure 4b shows the corresponding optical spectra at several selected frequencies. The overlay of single-sideband phase noise curves across the entire tuning range is shown in Figure 4c. Throughout this range, the phase noise at a 10 kHz offset from the carrier frequency remains below −130.02 dBc/Hz.

Despite its superior phase noise performance, the free-running OEO suffers from unsatisfactory low-frequency stability. As indicated by the black dash-dotted line in Figure 5a, it is difficult to identify a clear center frequency of the signal within a measurement span of 1 kHz. The phase noise at a 10 Hz offset, as depicted in Figure 5b, approaches 0 dBc/Hz; however, its reliability is compromised due to frequency drifting, making it difficult to discern accurately.

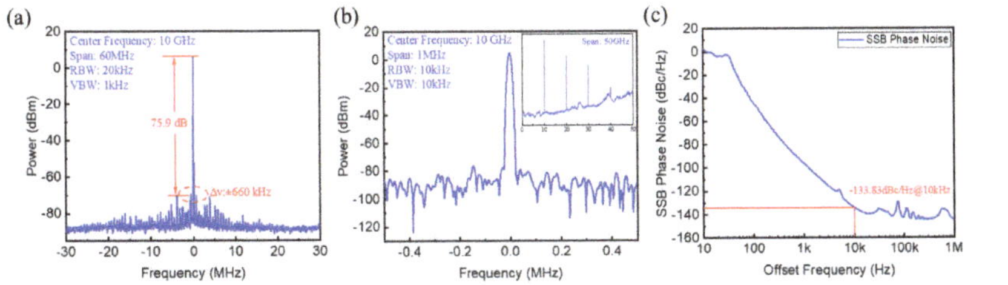

Figure 3. (**a**) RF spectrum with a span of 60 MHz; (**b**) RF spectrum with a span of 1 MHz; (**c**) single-sideband phase noise curve of a 10 GHz RF signal of free-running OEO.

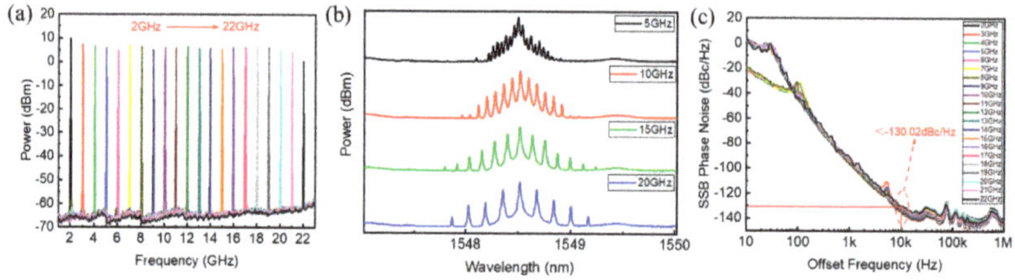

Figure 4. (**a**) RF Spectrum; (**b**) optical spectrum corresponding to individual frequencies; (**c**) single-sideband phase noise overlay of 2–22 GHz wideband-tuned RF signal of free-running OEO.

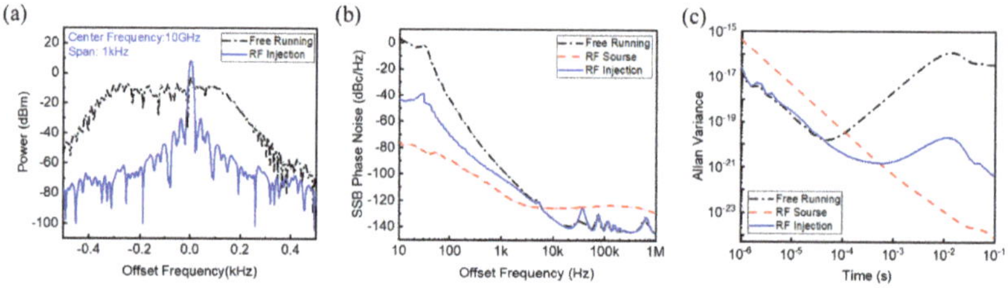

Figure 5. Comparison of 10 GHz RF signals between free-running OEO and injection-locked OEO: (**a**) RF spectrum at 1 kHz span; (**b**) phase noise; (**c**) Allan variance.

Subsequently, the OEO was injection-locked by its fundamental frequency, with the injection signal's phase noise measuring −75.87 dBc/Hz @ 10 Hz offset and −125.55 dBc/Hz @ 10 kHz offset, respectively. Following injection locking, a well-defined and stable RF tone (blue solid line) emerges within the identical measurement range. The phase noise decreases to −43.57 dBc/Hz @ 10 Hz (blue solid line), demonstrating a significant improvement of nearly 40 dB, as illustrated in Figure 5b. At frequencies far from the carrier, the phase noise remains close to that of the free-running OEO (e.g., −133.84 dBc/Hz @10 kHz), better than that of the injected fundamental RF signal. The comparison of Allan variance before and after injection locking is shown in Figure 5c. The Allan variance of the OEO decreases from 3.40×10^{-17} at 0.1 s in the free-running state to 3.99×10^{-22} at 0.1 s, representing an improvement of five orders of magnitude.

Next, the phase noise and Allan variance of the OEO under frequency tuning are evaluated. Figure 6 summarizes and compares the phase noise at a 10 kHz offset from the carrier frequency and the Allan variance at 0.1 s for the wideband-tunable OEO in both the free-running and fundamental injection-locking states. The phase noise remains nearly identical to that of the free-running OEO under RF injection, staying below −130 dBc/Hz, while the Allan variance improves by approximately five orders of magnitude across the entire tuning range compared to the free-running state.

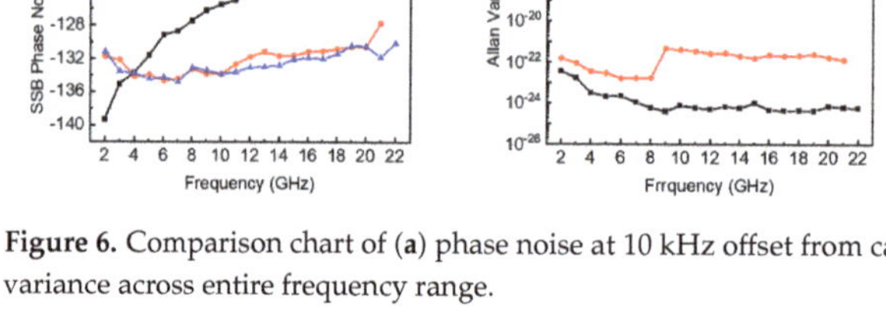

Figure 6. Comparison chart of (**a**) phase noise at 10 kHz offset from carrier frequency and (**b**) Allan variance across entire frequency range.

All the Allan variance measurements in this article were constrained to a maximum duration of 0.1 s due to the measurement limitations of the FSWP. To evaluate long-term frequency stability beyond this constraint, the output frequency drift of the OEO was characterized by recording its peak position over a 5 min period using the spectrum analyzer's MaxHold function, as shown in Figure 7. The resulting spectrum demonstrates a stable peak frequency and almost no frequency drift is observed within 5 min, which confirms the OEO system's long-term stability.

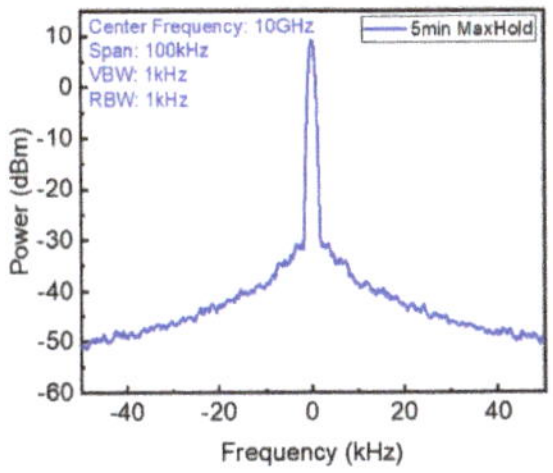

Figure 7. Five min MaxHold spectrum stability test curve of RF signal.

The influence of the injection power on the performance of the OEO is also investigated. Figure 8a,c depict the relationship between the phase noise and the Allan variance of the injection-locked OEO under different injection power levels. As the injection power increases from −24 dBm to 15 dBm, the phase noise @ 10 Hz decreases from −36.06 dBc/Hz to −58.40 dBc/Hz, showing a minimum 36 dB improvement compared to that of the free-running OEO. The phase noise far from the carrier frequency gradually converges to that of the free-running OEO. Figure 8b summarizes variation in the phase noise at a 10 Hz and 10 kHz offset from the carrier frequency with the injected RF signal power. Similarly, the improvement in Allan variance also increases with higher injection power, particularly for time intervals longer than 1 μs. Notably, for a time interval of 0.1 s, a minimum improvement of four orders of magnitude has been achieved, even with very weak injection. Figure 8d summarizes the variation in the Allan variance at 0.1 s under different injection powers. Investigations of the 15 GHz and 20 GHz cases also demonstrate similar performance.

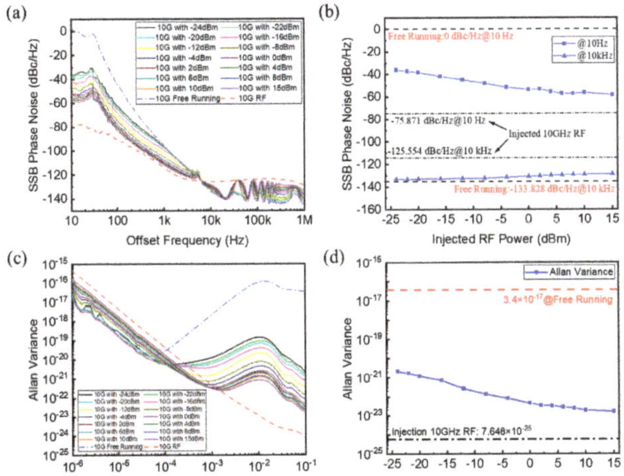

Figure 8. (**a**) Phase noise superposition curves of 10 GHz injection-locked OEO with different powers; (**b**) phase noise at 10 kHz offset from carrier frequency; (**c**) Allan variance superposition curves; and (**d**) Allan variance data corresponding to 0.1 s.

Finally, the minimum RF power required for the fundamental injection locking of the OEO was experimentally investigated. Measurements were taken at 2, 5, 10, 15, and 20 GHz, as shown in Figure 9a. Across the entire tuning range of the OEO, the minimum power required for fundamental injection locking gradually rises from −50 dBm @ 2 GHz to −18 dBm @ 20 GHz due to degradation of the system response as frequency increases. Figure 9b,c show the phase noise and Allan variance of the OEO system in the minimum-RF-power injection-locking state, respectively. Notably, across the entire tuning range, the phase noise of the OEO system remains the same as that in the free-running state, while the Allan variance is significantly improved, consistent with the earlier experimental findings.

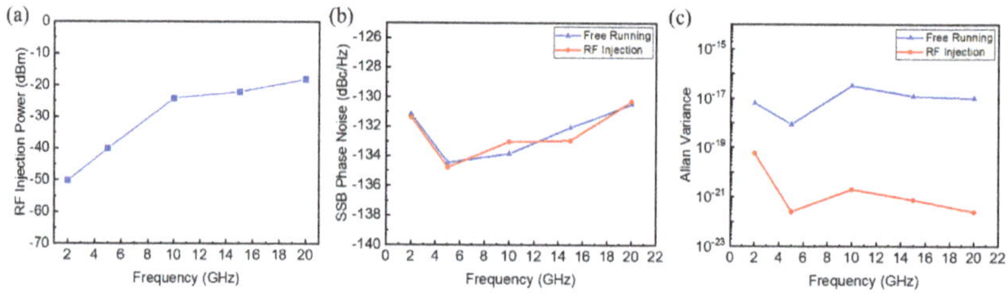

Figure 9. Comparison curves of (**a**) minimum RF power required; (**b**) phase noise and (**c**) Allan variance of the OEO system in free-running state and in minimum locked power injection state at different frequencies (corresponding to 2/5/10/15/20 GHz, respectively).

3.2. Subharmonic RF Signal Injection Locking

Fundamental injection locking is effective in reducing close-to-carrier phase noise and improving the Allan variance of the OEO. To further reduce system costs associated with the wideband synthesizer, a subharmonic injection-locking scheme is explored. This approach allows a synthesizer operating at the $1/n$ fundamental frequency of the OEO to achieve the desired frequency tuning range. For the OEO system, the n-th harmonic of the injected RF signal can be directly obtained from the Mach-Zehnder modulator, using its inherent nonlinear response. Based on the previous investigation, which demonstrated that the minimum injection power for fundamental injection locking can be reduced to −20 dBm, it is feasible to obtain the n-th harmonic of the subharmonic injection signal with sufficient power to effectively lock the OEO.

The subharmonic injection-locking range can be expressed as [14]:

$$\Delta\omega_{\frac{1}{n}} = \frac{\omega_0}{2Q}\frac{U_{in}}{U_{out}} = \frac{\omega_0}{2Q}\sqrt{\frac{P_{in}}{P_{out}}} \quad (4)$$

The locking range is also related to the Q factor and the injection power of the injected signal.

The OEO was first investigated using the 1/2, 1/3, 1/4, and 1/5 subharmonics of 10 GHz, each with a power of −7 dBm. Figure 10 presents the phase noise and Allan variance of the signal generated by the OEO under different subharmonic injection-locking conditions. As shown in Figure 10a, regardless of the injected subharmonic frequency, the phase noise at a 10 Hz offset from the carrier frequency is reduced by approximately 40 dB, and the phase noise curves of the subharmonic injection-locked OEO system remain highly consistent within a 1 kHz offset from the carrier frequency. Beyond 1 kHz, however, the phase noise of the OEO system increases after subharmonic injection locking. As n. increases, the phase noise curve at frequencies farther from the carrier also rises.

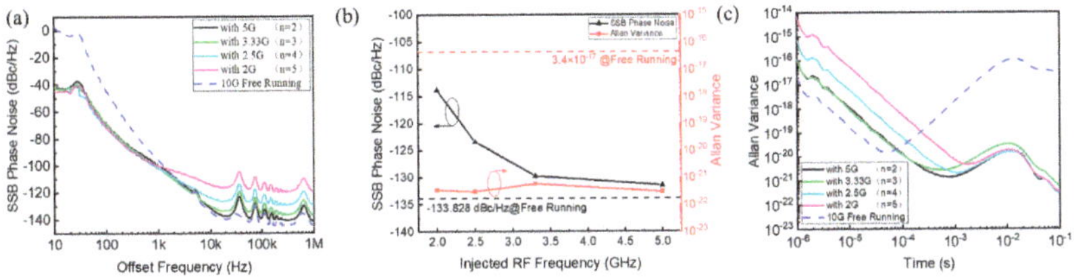

Figure 10. (**a**) Comparison curve of phase noise; (**b**) Summary of phase noise at 10 kHz and Allan variance corresponding to 0.1 s; (**c**) Comparison curve of Allan variance of the OEO system in the locked state under the same power of 1/2, 1/3, 1/4, and 1/5 subharmonic injection.

Figure 10b summarizes the phase noise at a 10 kHz offset from the carrier frequency under different injection orders. The phase noise gradually deteriorates, increasing from −133.83 dBc/Hz in the free-running state to −131.49 dBc/Hz @ 5 GHz injection, −129.77 dBc/Hz @ 3.3 GHz injection, −123.43 dBc/Hz @ 2.5 GHz injection, and −113.87 dBc/Hz @ 2 GHz injection, respectively. This degradation occurs because, as n increases, the subharmonic signal becomes weaker under the same modulation depth. The practical subharmonic order depends on the extent to which the phase noise degrades to outperform that of the fundamental RF signal. For the 10 GHz case, our RF synthesizer can generate a signal with a phase noise of −125.55 dBc/Hz at 10 GHz. Based on this, the effective subharmonic order is determined to be 3, corresponding to a phase noise of −129.77 dBc/Hz at 3.3 GHz. For an RF synthesizer with poorer phase noise performance, the usable subharmonic order can be further increased if the goal is to obtain an RF signal with better phase noise characteristics than the synthesizer itself.

Figure 10c illustrates the Allan variance of the OEO within the 0.1 s measurement range before and after subharmonic injection locking. In the free-running state, the Allan variance of the OEO system at 0.1 s is 3.4×10^{-17}. After subharmonic injection, the Allan variance for all cases at 0.1 s decreases to approximately 5×10^{-22}, representing a reduction of nearly five orders of magnitude, similar to the improvement achieved with fundamental injection locking.

Next, the influence of subharmonic power on phase noise and Allan variance is investigated using 1/3 subharmonic injection. As shown in Figure 11a, when the power of the injected subharmonic signal increases from −7 dBm to 10 dBm, the phase noise at a 10 Hz offset from the carrier frequency improves from −42.05 dBc/Hz @ −7 dBm to −59.63 dBc/Hz @ 10 dBm, achieving a maximum reduction of approximately 40 dB compared to the free-running case. At a 10 kHz offset from the carrier frequency, the phase noise remains relatively stable, ranging from −126.68 dBc/Hz to −129.77 dBc/Hz, which represents a 5~7 dB decrease compared to that of the fundamental injection locking. Figure 11b summarizes the phase noise at 10 Hz and 10 kHz offsets for different injection power levels.

Figure 11c,d illustrate the influence of subharmonic power on the Allan variance. As the injection power increases, the Allan variance at 0.1 s improves from 5.87×10^{-22} at −7 dBm to 1.65×10^{-23} at 10 dBm, demonstrating a dependence on subharmonic injection power. However, for all injection power levels, the Allan variance can be optimized to five orders of magnitude.

The minimum RF power required to lock the OEO at different subharmonic orders is investigated. We define the critical injection-locking state as the point where the Allan variance of the injected OEO becomes comparable to that of the free-running OEO. The results are presented in Figure 12. For a specific frequency, when injecting $1/n$ subharmonics of different frequencies, the minimum RF power required for the OEO to reach the locked state increases as n increases. Additionally, when n is constant, the minimum RF power required for locking rises as the free-running frequency of the OEO increases. This trend is consistent with that observed in the fundamental injection-locked OEO system.

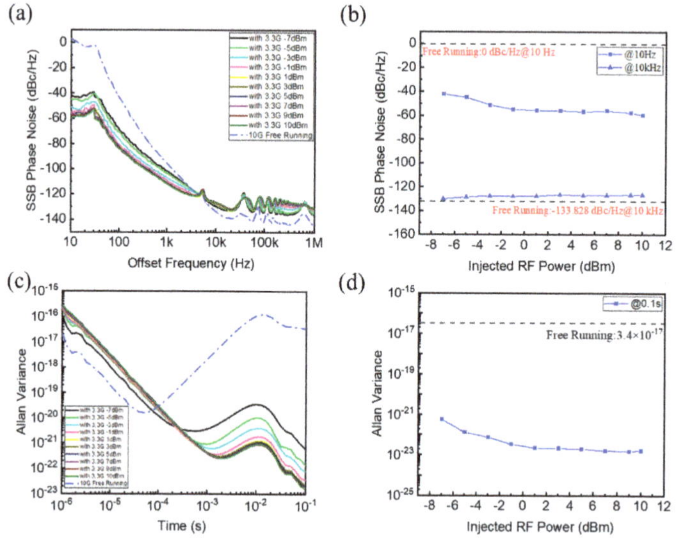

Figure 11. (**a**) Phase noise superposition curves of 10G OEO in locked state of 1/3 harmonic RF injection at different powers; (**b**) phase noise at 10 Hz and 10 kHz offset from carrier frequency; (**c**) Allan variance superposition curves; and (**d**) Allan variance data summary corresponding to 0.1 s.

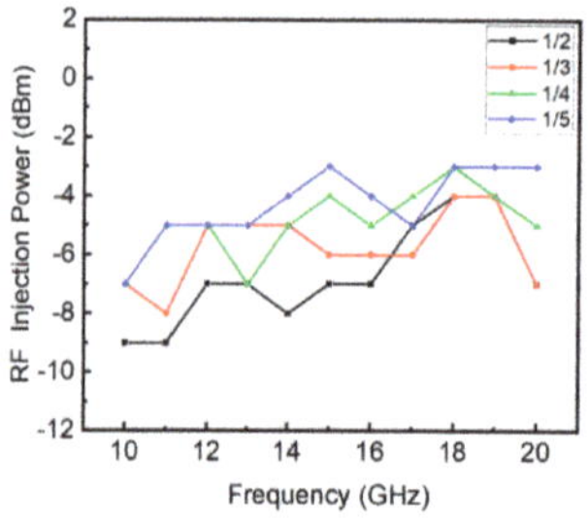

Figure 12. The variation curve of the minimum injected power required for each frequency to achieve the locking state at subharmonic injections at different frequencies.

4. Conclusions

This paper presents a wideband-tunable, low-phase-noise OEO based on fundamental and subharmonic injection locking. The injection-locked OEO enables a 2–22 GHz wideband-tunable RF signal output, featuring high spectral purity, a high spurious suppression ratio, and low phase noise. Phase noise levels below −130.02 dBc/Hz at a 10 kHz offset have been demonstrated across the full frequency tuning range for the fundamental injection-locking configuration. The close-in phase noise of the OEO-generated RF signal is reduced by approximately 40 dB at a 10 Hz offset from the carrier frequency, while the maintaining low-phase-noise characteristics of the OEO at frequencies several kHz from the carrier. The Allan variance improves by about five orders of magnitude at a 0.1 s. Subharmonic injection locking is also demonstrated for 10 GHz RF generation, with phase noise performance degrading from −131.49 dBc/Hz to −113.87 dBc/Hz at a 10 kHz offset as the subharmonic order (n) increases from 2 to 5. The influence of injection power on the system performance is also examined.

This scheme facilitates the easy implementation of frequency stabilization for high-performance OEOs. Depending on specific requirements, either the fundamental or subharmonic injection schemes can be employed to achieve high-performance or cost-effective RF generation. Furthermore, this work offers a technical reference for optimizing the stability performance of high-performance wideband-tunable opto-generated RF sources.

Author Contributions: Conceptualization, D.L. and Z.Z.; methodology, Z.Z.; formal analysis, Z.Z. and H.S.; investigation, Z.Z.; resources, D.L. and L.Z.; data curation, Z.Z.; writing—original draft preparation, Z.Z.; writing—review and editing, D.L, F.G. and Z.Z.; visualization, Z.Z.; supervision, D.L. and L.Z.; funding acquisition, D.L. All authors have read and agreed to the published version of the manuscript.

Funding: This work was supported by the National Key Research and Development Program of China under Grant No. 2020YFB1805700.

Institutional Review Board Statement: Not applicable.

Informed Consent Statement: Not applicable.

Data Availability Statement: The original contributions presented in this study are included in the article. Further inquiries can be directed to the corresponding author.

Conflicts of Interest: The authors declare no conflicts of interest.

References

1. Yao, X.S.; Maleki, L. Optoelectronic Microwave Oscillator. *J. Opt. Soc. Am. B* **1996**, *13*, 1725–1734. [CrossRef]
2. Lu, H.-H.; Li, C.-Y.; Lu, T.-C.; Wu, C.-J.; Chu, C.-A.; Shiva, A.; Mochii, T. Bidirectional Fiber-Wireless and Fiber-VLLC Transmission System Based on an OEO-Based BLS and a RSOA. *Opt. Lett.* **2016**, *41*, 476–479. [CrossRef] [PubMed]
3. Xu, L.; Hu, X.; Zhang, Y.; Yi, J.; Yu, Y.; Xiao, X.; Yu, Y. A Highly Sensitive and Precise Temperature Sensor Based on Optoelectronic Oscillator. *Opt. Commun.* **2021**, *483*, 126625. [CrossRef]
4. Wang, J.; Yu, J.; Miao, W.; Sun, B.; Jia, S.; Wang, W.; Wu, Q. Long-Range, High-Precision Absolute Distance Measurement Based on Two Optoelectronic Oscillators. *Opt. Lett.* **2014**, *39*, 4412–4415. [CrossRef] [PubMed]
5. Lou, C.; Huo, L.; Chang, G.; Gao, Y.Z. Experimental Study of Clock Division Using the Optoelectronic Oscillator. *IEEE Photon. Technol. Lett.* **2002**, *14*, 1178–1180. [CrossRef]
6. Yao, X.S.; Maleki, L. Multiloop Optoelectronic Oscillator. *IEEE J. Quantum Electron.* **2000**, *36*, 79–84. [CrossRef]
7. Tang, H.; Yu, Y.; Wang, Z.; Xu, L.; Zhang, X. Wideband Tunable Optoelectronic Oscillator Based on a Microwave Photonic Filter with an Ultra-Narrow Passband. *Opt. Lett.* **2018**, *43*, 2328–2331. [CrossRef] [PubMed]
8. Zhang, Y.; Hou, D.; Zhao, J. Long-Term Frequency Stabilization of an Optoelectronic Oscillator Using Phase-Locked Loop. *J. Light. Technol.* **2014**, *32*, 2408–2414. [CrossRef]
9. Zhang, J.; Yao, J. Parity-Time–Symmetric Optoelectronic Oscillator. *Sci. Adv.* **2018**, *4*, eaar6782. [CrossRef] [PubMed]
10. Fan, Z.; Su, J.; Lin, Y.; Jiang, D.; Chen, Y.; Li, X.; Qiu, Q. Injection Locking and Pulling Phenomena in an Optoelectronic Oscillator. *Opt. Express* **2021**, *29*, 4681–4699. [CrossRef] [PubMed]
11. Sarkar, J.; Banerjee, A.; Pacheco, G.M.; Das, N.R. Experimental Investigation on the Side Mode Injection Locking of a Single-Loop OEO under RF Signal Injection. *Analog. Integr. Circuits Signal Process.* **2024**, *118*, 539–552. [CrossRef]
12. Peng, D.; Hong, W.; Xiao, H.; Han, Y.; Huang, Q.; Qiu, S.; Li, J.; Xiang, M.; Fu, S.; Qin, Y. Freely Tunable Phase-Locked Dual-Frequency Microwave Signal Generation from Injection-Locked Broadband Optoelectronic Oscillator. *Opt. Express* **2024**, *32*, 41857–41869. [CrossRef] [PubMed]
13. Banerjee, A.; Aguiar Dantas De Britto, L.; Mendes Pacheco, G. Analysis of Injection Locking and Pulling in Single-Loop Optoelectronic Oscillator. *IEEE Trans. Microw. Theory Technol.* **2019**, *67*, 2087–2094. [CrossRef]
14. Daryoush, A.S.; Berceli, T.; Saedi, R.; Herczfeld, P.R.; Rosen, A. Theory of Subharmonic Synchronization of Nonlinear Oscillators. In Proceedings of the IEEE MTT-S International Microwave Symposium Digest, Long Beach, CA, USA, 13–15 June 1989; pp. 735–738.

photonics

MDPI

Article

Ultra-Compact Multimode Micro-Racetrack Resonator Based on Cubic Spline Curves

Zhen Li [1,2], Chuang Cheng [1,2], Xin Fu [1] and Lin Yang [1,2,*]

[1] State Key Laboratory of Optoelectronic Materials and Devices, Institute of Semiconductors, Chinese Academy of Sciences, Beijing 100083, China; lizhen2019@semi.ac.cn (Z.L.); chengchuang@semi.ac.cn (C.C.); fuxin@semi.ac.cn (X.F.)

[2] College of Materials Science and Opto-Electronic Technology, University of Chinese Academy of Sciences, Beijing 100049, China

* Correspondence: oip@semi.ac.cn

Abstract: Micro-racetrack resonators have become one of the key components for realizing signal processing, generation, and integration in microwave photonics, owing to their high Q factor, compact footprint, and tunability. However, most of the reported micro-racetrack resonators are confined to the single-mode regime. In this paper, we designed an ultra-compact multimode micro-racetrack resonator (MMRR) based on shape-optimized multimode waveguide bends (MWBs). Cubic spline curves were used to represent the MWB boundary and adjoint methods were utilized for inverse optimization, achieving an effective radius of 8 μm. Asymmetric directional couplers (ADCs) were designed to independently couple three modes into a multimode micro-racetrack, according to phase-matching conditions and transmission analysis. The MMRR was successfully fabricated on a commercial platform using a 193 nm dry lithography process. The device exhibited high loaded Q factors of 2.3×10^5, 4.1×10^4, and 2.9×10^4, and large free spectral ranges (FSRs) of 5.4, 4.7, and 4.2 nm for TE_0, TE_1, and TE_2 modes, with about a 19×55 μm^2 footprint.

Keywords: micro-racetrack resonator; inverse design; multimode waveguide bend; silicon photonics; microwave photonics

1. Introduction

Currently, communication systems face significant challenges in terms of energy consumption and bandwidth, and microwave photonics is an important technological solution to address these challenges [1]. However, traditional microwave photonic systems still suffer from large size, poor stability, and high power consumption. The application of photonic integration technology can not only greatly reduce the system size and power consumption but can also significantly enhance its performance, holding the promise of realizing a single microwave photonic processing chip [2]. Micro-racetrack resonators, one of the key components in integrated optics, have demonstrated excellent performance in high speed modulators [3,4], optical switches [5,6], microwave photonic filters [7,8], and many other fields. Meanwhile, to enhance the communication capacity of optical networks, the utilization of multiplexing technology has emerged as an effective strategy. Combined with silicon-based optoelectronic technology, a variety of chip-scale multiplexing devices and systems have been reported [9–11]. Mode division multiplexing (MDM) devices enable simultaneous parallel data transfers on a single physical channel, having attracted widespread attention [12,13]. Recently, numerous multimode devices were demonstrated for silicon-based MDM systems, including mode (de)multiplexers [14,15], multimode

https://doi.org/10.3390/photonics12040326

waveguide crossings [16,17], multimode power splitters [18], and multimode fiber-to-chip couplers [19,20].

MWB is a vital device of MDM technologies [13,21]. However, traditional bends require bending radii of several hundred millimeters [22]. Numerous schemes have been reported to reduce the footprint of MWBs. In 2017, particle swarm optimization (PSO) was introduced to achieve a MWB with an effective radius of 5 μm [23], whereas, it could only support two TE modes. H. Wu et al. presented a grating-like structure with a radius of 10 μm using subwavelength grating [24], but precise lithographic control is needed due to the tiny gaps. Special mathematical curves, such as free-form curve (FFC) and Bezier curve, were also employed to achieve compact MWBs [25–28].

Most of the micro-racetrack resonators are designed to operate in the fundamental mode case, limited by mode crosstalk and mode dispersion. Separately optimizing micro-racetrack resonator parameters for each mode is clearly effective for multimode multiplexing [29]; however, multiple single mode micro-racetracks (MSMMR) will increase the overall size of the device. In [30], a shape-optimized MWB based on transformation optics (TO) is demonstrated and applied into a MMRR, whereas the effective radius of the MWB is 15 μm. In [31], a MMRR composed of modified Euler waveguide bends and a subwavelength grating (SWG) coupler is designed, but the MMRR exhibits high insertion loss (IL) and mode crosstalk. Meanwhile, small feature sizes of the SWG coupler may not be compatible with standard silicon photonic fabrication processes. Euler waveguide bends are also applied to the lithium-niobate-on-insulator (LNOI) platform [32]. Whereas this MMRR can only be operated at TE_0 mode. In 2024, a taper waveguide is introduced to achieve an AlGaAs-on-insulator (AlGaAsOI) MMRR [33]. Another novel MMRR is proposed and demonstrated in [34], with the help of total internal reflection (TIR) effect. However, the insertion loss of this MMRR is also large due to the scattering at the TIR mirror surface.

In this work, we propose and demonstrate an ultra-compact and low-loss MMRR on a commercial 220 nm silicon-on-insulator (SOI) platform. The MMRR consists of optimized MWBs and multimode straight waveguides, supporting the first three-order modes (TE_0, TE_1, and TE_2). The shape of the MWB is described by cubic spline curves and is inverse designed and optimized by adjoint methods. The proposed MWB, which is also used as a bending directional coupler for the TE_0 mode, possesses an effective radius of 8 μm. The TE_1 and TE_2 modes are coupled into MMRR through the corresponding ADC, and the total length of the MMRR is about 149 μm. The fabricated MMRR exhibits > 2.9×10^4 loaded Q factors and about 5 nm FSR for the three modes.

2. Design and Simulation

2.1. The Design of Multimode Waveguide Bend (MWB)

Conventional MWBs usually require a very large bending radius to avoid undesired losses and inter-mode crosstalk due to mode mismatch [30,35]. The performance of a MWB is strongly dependent on its shape, and several different special mathematical curves have been employed to optimize the MWB [36–38]. Among them, the cubic spline interpolation curves possess a high degree of freedom and are not limited by the well-defined curve functions [38]. For a given set of n data points (x_i, y_i) that follow $x_1 < x_2 < \ldots < x_n$, a smooth curve can be obtained by cubic spline interpolation, in which the adjacent data points are fitted by a cubic polynomial:

$$S_i(x) = y_i + b_i(x - x_i) + c_i(x - x_i)^2 + d_i(x - x_i)^3, \quad x_i \leq x \leq x_{i+1}. \tag{1}$$

where $i = 1, 2, \ldots, n-1$, $S_i(x)$ is the *i*th spline curve, b_i, c_i, and d_i are unknown coefficients. Cubic spline curves require the segments to be continuous at the nodes, and the first-order and second-order derivatives to be continuous at the nodes, which can be expressed as follows:

$$\begin{aligned} y_i &= S_{i-1}(x_i), i = 2,3,\ldots,n. \\ S_i'(x_{i+1}) &= S_{i+1}'(x_{i+1}), i = 1,2,\ldots,n-2. \\ S_i''(x_{i+1}) &= S_{i+1}''(x_{i+1}), i = 1,2,\ldots,n-2. \end{aligned} \tag{2}$$

the total number of unknown coefficients is $3n - 3$ from Equation (1), while only $3n - 5$ equations can be formulated according to Equation (2). Thus, it is often necessary to impose additional boundary conditions at the start and end points of the spline curve in practical applications. A commonly employed natural boundary condition can be formulated as follows [38]:

$$\begin{aligned} S_1''(x_1) &= 0, \\ S_{n-1}''(x_n) &= 0. \end{aligned} \tag{3}$$

by combining Equation (1) to Equation (3), we can solve all the undetermined coefficients.

As shown in Figure 1a, to balance the trade-off between device size and performance, we choose the effective radius of the MWB as 8 μm. The width of both ends of the MWB is 1.3 μm, consistent with the straight multimode waveguide, supporting the first three-order TE modes. The MWB consists of two identical 45-degree curved sections, with its boundary outlined by two sets of cubic spline interpolation points, which are marked as blue circles. To obtain accurate device performance and smooth curve shapes, we choose 20 and 15 interpolation points in the inner and outer curves of the MWB. The optimization of this MWB by using traditional parameter scanning methods requires a large amount of computational resources and time. Therefore, we employ the adjoint method to optimize the dataset, which is significantly efficient for the inverse design of complex electromagnetic components [39,40]. By calculating the gradient of figure-of-merit (FOM) through only two simulations in each iteration, namely, forward and adjoint simulations, the adjoint method can significantly accelerate the optimization process.

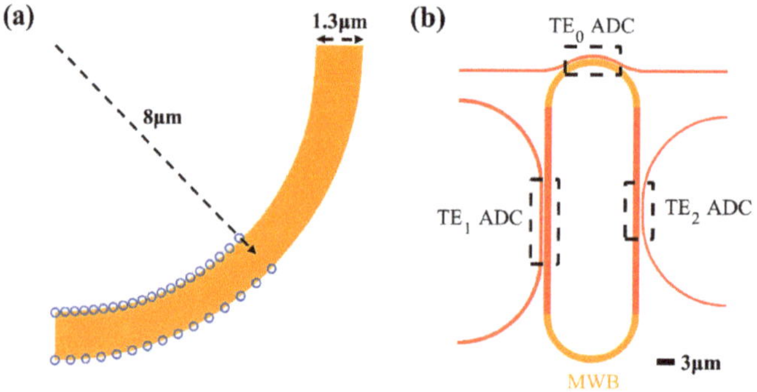

Figure 1. (**a**) Schematic of the MWB with an effective radius of 8 μm and a waveguide width of 1.3 μm. The control points of the cubic spline curve are denoted as blue circles. (**b**) Diagram of the proposed MMRR. The TE_0, TE_1, and TE_2 ADCs are highlighted by the black dashed box.

The FOM is defined as mode overlap (η) between the input mode and the corresponding output mode of the MWB, similar to formulations in [35,41], and is easy to implement in Lumerical softwares from Ansys Canada Ltd. (Canonsburg, PA, Canada) The total FOM can be written as follows:

$$\mathrm{FOM} = \prod_{i=0}^{2} \mathrm{FOM_i} = \prod_{i=0}^{2} \eta_i \tag{4}$$

where i denotes the ith mode. A 1/4 ring waveguide with a radius of 8 μm and a width of 1.3 μm is chosen as the initial structure, and the interpolation points are uniformly distributed. Aided by the shape optimization within the Lumopt module of Ansys Lumerical, the 3D finite-difference time-domain (FDTD) solver is selected to maximize the FOM, and the L-BFGS-B algorithm is used to update the parameters [42]. The structure of the optimized MWB approximates a 1/4 circle that is wider at both ends and narrower in the middle. Figure 2 illustrates the optical field distributions of the three modes through the MWB and the corresponding insertion loss and crosstalk curves. There is no significant mode mismatch between the input and output waveguide ports. Moreover, within the wavelength range of 1.5 to 1.6 μm, the average losses for the TE_0, TE_1, and TE_2 modes are 0.01 dB, 0.03 dB, and 0.03 dB, and the crosstalks are below −21.3 dB, −17.9 dB, and −17.3 dB, respectively.

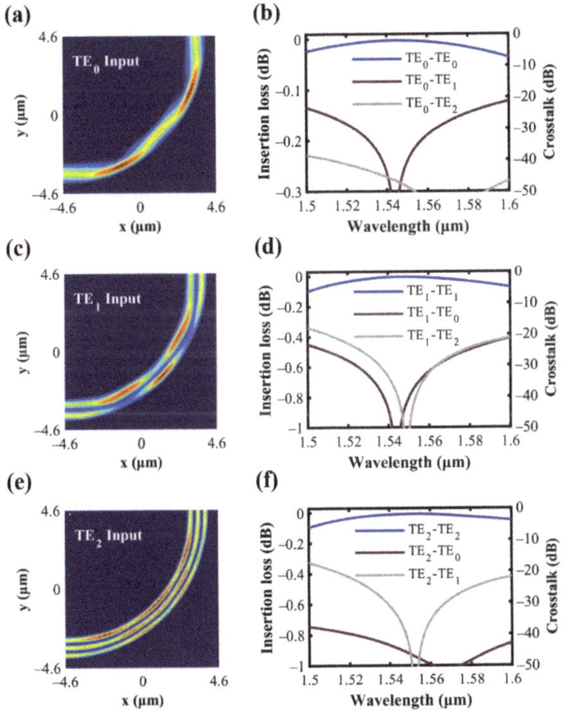

Figure 2. The z component of the magnetic field H_z distribution and transmission spectra for (**a**,**b**) TE_0, (**c**,**d**) TE_1, and (**e**,**f**) TE_2 modes. The average insertion losses for three modes are 0.01 dB, 0.03 dB, and 0.03 dB, respectively. The maximum crosstalk for TE_0, TE_1, and TE_2 modes are −21.3 dB, −17.9 dB, and −17.3 dB, respectively.

2.2. The Design of Multimode Micro-Racetrack Resonator (MMRR)

The proposed MMRR consists of a multimode micro-racetrack located at the center and three ADCs positioned at the periphery, as shown in Figure 1b. According to the coupled mode theory [43], when two waveguides are close enough, there will be periodic energy exchange between them via the evanescent fields. The employed ADC consists of a single-mode waveguide and a multimode waveguide, which enables mode conversion between a fundamental mode and a high-order mode. To achieve efficient mode conversions, the three coupling regions should be designed carefully, including widths of access waveguides, coupling lengths, and gaps between waveguides, as shown in Figure 3.

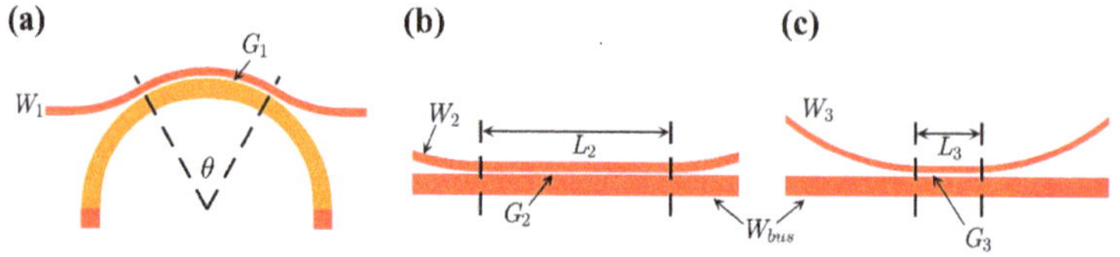

Figure 3. (**a**) Detailed schematic for bend coupling region of TE_0 ADC; the outer boundary of the MWB is set to 1/4 arc to guarantee the performance. Diagram of straight coupling region for (**b**) TE_1 ADC (**c**) TE_2 ADC; these parameters are chosen carefully to obtain the desired low crosstalk and transmission coefficient.

Figure 4a shows the effective indices of the first three-order modes in the waveguides with different widths at the wavelength of 1.55 µm, calculated by the finite difference eigenmode method. For the TE_1 and TE_2 modes, $n_{eff}^1 = n_{eff}^2$ should be fulfilled according to the phase-matching condition, where n_{eff}^1 is the effective indices for the TE_0 modes of the two access waveguides, and n_{eff}^2 is the effective indices for the TE_1 or TE_2 modes of the multimode waveguide. Based on this principle, we choose the widths of the access waveguides (W_2 and W_3) as 0.64 µm and 0.41 µm, respectively. Moreover, a bending ADC is used to couple the TE_0 mode of the narrow access waveguide into the TE_0 mode of the MWB. Similarly, $n_{eff}^1 R_1 = n_{eff}^2 R_2$ should be fulfilled according to the phase-matching condition [44], where R_1 and R_2 are the radii of curvature of the two bending waveguides. The width W_1 of the TE_0 ADC can be calculated as 0.52 µm according to this formula.

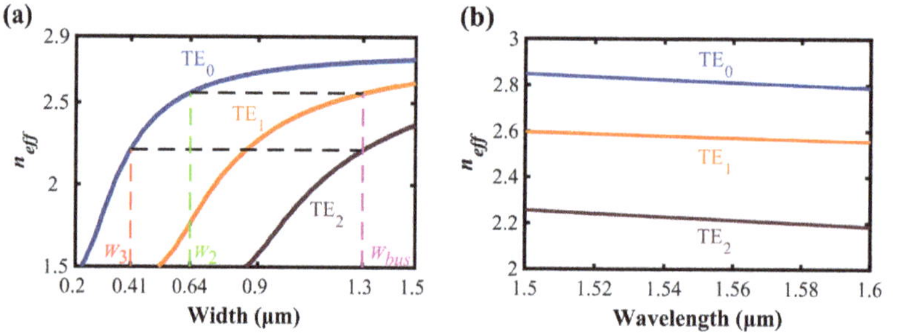

Figure 4. Effective indices of TE_0, TE_1, and TE_2 modes for different (**a**) widths (for the calculation of W_2 and W_3 which are denoted as green and red) and (**b**) wavelengths (for the calculation of the group refractive index).

Coupling lengths and gaps between waveguides affect the coupling coefficients of the MMRR, which determine the MMRR performance, like the IL and extinction ratio (ER). To make the MMRR suitable for integrated optoelectronic applications, a large FSR and ER are usually required. Thus, the expected FSR and ER of the MMRR are set to 5 nm and 20 dB. The FSR of an all-pass micro-racetrack resonator can be expressed as below:

$$FSR = \frac{{\lambda_c}^2}{n_g L} \tag{5}$$

where λ_c = 1.55 µm, n_g is the group refractive index of the mode, and L is the circumference of the multimode micro-racetrack. n_g is defined as follows:

$$n_g = n_{eff} - \lambda \frac{dn_{eff}}{d\lambda} \tag{6}$$

Figure 4b demonstrates the effective refractive index of each mode as a function of wavelength. It is fitted to a straight line using the least-squares method and the slope of the line to be brought into Equation (6) is derived to obtain the n_g of each mode. We first consider the TE_1 mode in the multimode waveguide, which has a n_g of 3.2325 at wavelength of 1.55 μm; so, the circumference L can be derived from Equation (5) as 148.65 μm under the target of an FSR of 5 nm.

To choose the coupling coefficients t of the three ADCs, we consider the relationship between the self-coupling coefficient and ER in the all-pass micro-racetrack resonator, which can be expressed as follows [45]:

$$ER = -20log\left(\frac{(t+\alpha)(1-\alpha t)}{(t-\alpha)(1+\alpha t)}\right) \quad (7)$$

where α is the attenuation factor of the multimode micro-racetrack resonator. For the ring SOI waveguide with 5–10 μm bending radius, the loss is generally considered to be 5–20 dB/cm, which corresponds to an attenuation factor α of 0.98–0.99. Considering the inevitable fabrication errors, we analyzed three representative values of α, namely 0.98, 0.985, and 0.99, by which t can be calculated according to Equation (7).

Once the coupling coefficients t are clear, we can choose the appropriate coupling length by 3D FDTD simulations. To reduce the complexity of the device fabrication while simultaneously maintaining a compact device size, we selected the gaps of the three ADCs to be 200 nm, 200 nm, and 300 nm, respectively. Taking the case of α = 0.985 as an example, by carrying out 3D FDTD parameter scanning, we can derive the corresponding coupling region angle and the lengths of the three ADCs as 51.4 degrees, 10.57 μm, and 2.3 μm at the wavelength of 1.55 μm. Figure 5 illustrates the optical field distributions for the launched TE_0, TE_1, and TE_2 modes in the three ADCs. In the TE_0 ADC, the performance of the bending coupling is guaranteed due to the refractive index matching between the modes and the fact that the interpolation points at the outer boundary of the MWB are almost distributed along a conventional 1/4 bend. In the other two ADCs, the TE_1 and TE_2 modes injected from the bus multimode waveguide are coupled to the corresponding TE_0 modes of the access waveguides as expected. No mode crosstalk is observed in each ADC.

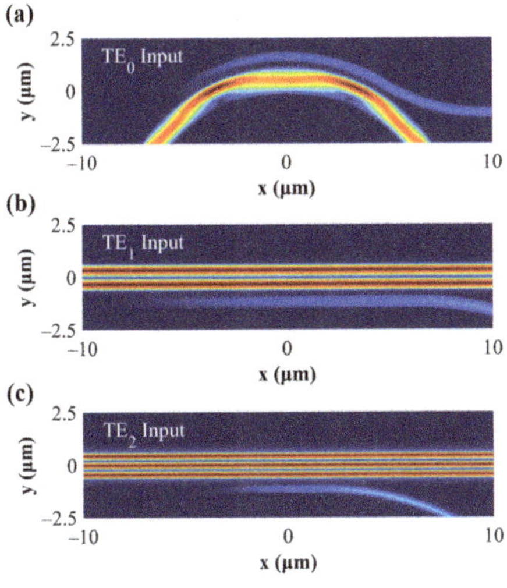

Figure 5. (**a**) TE_0 ADC, (**b**) TE_1 ADC, and (**c**) TE_2 ADC electric field distributions for the MMRR at 1.55 μm.

Some of the design parameters for the three ADCs of the MMRR are listed in Table 1, in which α is a dimensionless parameter. In addition, other parameters can be summarized as follows: W_1 = 0.52 μm, W_2 = 0.64 μm, W_3 = 0.41 μm, W_{bus} = 1.3 μm, G_1 = 0.2 μm, G_2 = 0.2 μm, G_3 = 0.3 μm.

Table 1. Design parameters for three ADCs of the MMRR.

α	θ (Degree)	L_2 (μm)	L_3 (μm)
0.98	62.57	12.74	3.27
0.985	51.4	10.57	2.3
0.99	42.86	8.03	1.28

We also calculated the coupling coefficient t at different wavelengths when the above parameters are employed, using the 3D FDTD solver. As illustrated in Figure 6, in the wavelength range of 1.5 to 1.6 μm, t is higher than 0.97 and increases with α. Since the coupling coefficient t is no longer linearly related to the wavelength, we can use polynomial fitting to derive the corresponding mathematical expression.

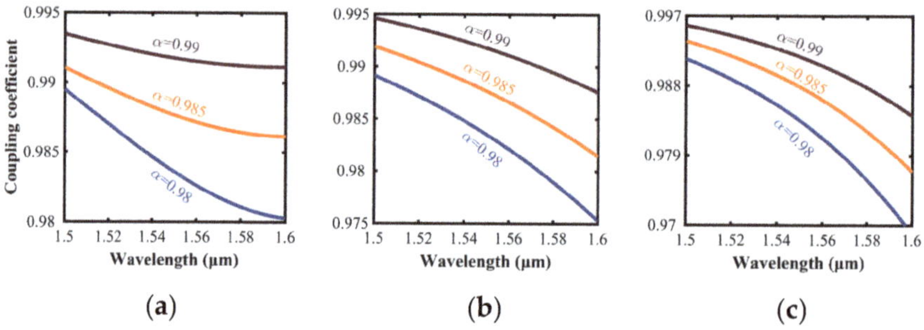

Figure 6. The relationship between coupling coefficient and wavelength for (**a**) TE_0 ADC, (**b**) TE_1 ADC, and (**c**) TE_2 ADC under different attenuation factors.

The transmission spectra of the three modes in the MMRR are then calculated. The relationship between transmission efficiency (T) and the wavelength of an all-pass micro-racetrack resonator can be expressed as follows [45]:

$$T = 10log\left(\frac{\alpha^2 + t^2 - 2\alpha t cos\theta}{1 + \alpha^2 t^2 - 2\alpha t cos\theta}\right) \tag{8}$$

where $\theta = 2\pi n_{eff} L/\lambda$, which is the phase of the optical wave after propagating one complete round trip in the micro-racetrack resonator. Simultaneously considering the dispersion characteristics of n_{eff} and t, shown in Figures 4b and 6, the theoretical transmission spectrum for each mode can be calculated according to Equation (8).

As shown in Figure 7, under these different α, the FSRs of the three modes are 5.2 nm, 5 nm, and 5.9 nm, respectively. Here, we need to point out that since we first determine the circumference of the MMRR based on the n_g of the TE_1 mode with (5), and this value will not be changed for a determined MMRR, this leads to a slight difference in the FSR due to the n_g distinction of the three modes.

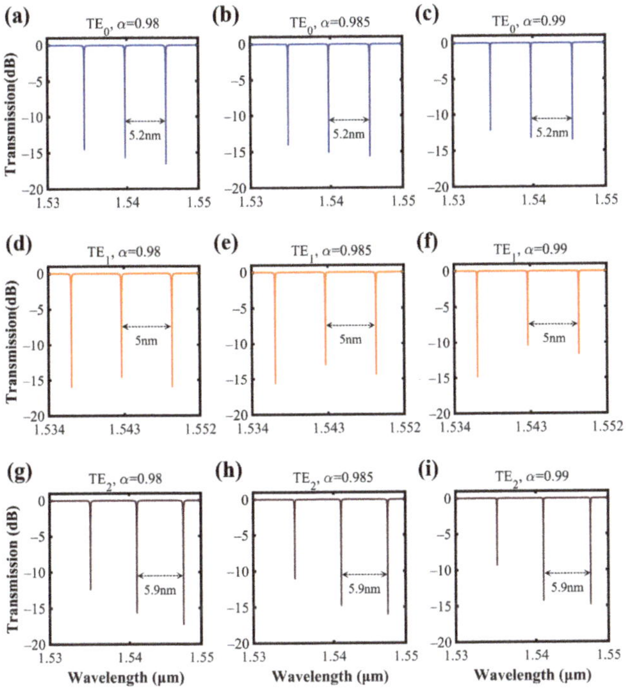

Figure 7. Transmission spectra of the MMRR for different attenuation factors and modes.

3. Fabrication and Characterization

Using a 193 nm dry lithography process, the designed MMRRs were fabricated on a 220 nm-thick SOI wafer with a 3 μm-thick silicon dioxide buried layer at Advanced Micro Foundry, Singapore. Figure 8 presents the microscope image of the devices, with clear boundaries and smooth contours shown in the etched MMRR. The footprint of the MMRR is only ~19 × 55 μm^2, and six identical grating couplers are used for optical transmission between the device and the single-mode fibers.

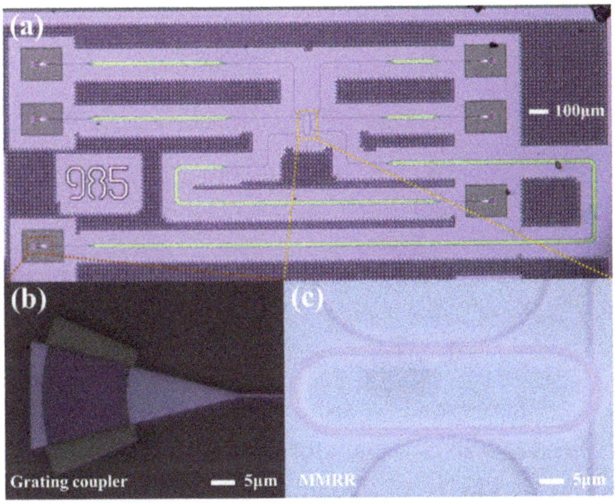

Figure 8. Microscope images of the fabricated MMRR. (**a**) Overall picture for fabricated devices. (**b**) Grating coupler. (**c**) The zooming-in micrographs of MMRR.

An amplified spontaneous emission (ASE) source and an optical spectrum analyzer (OSA) are used to measure the transmission spectra of each mode in the fabricated MMRR. The ASE has a flat spectrum to support the measurement. To accurately characterize the spectra, the wavelength resolution of OSA is set to be 10 pm. The grating coupler is calibrated to obtain the best tilt angle and the lowest transmission loss. The transmission spectrum of the reference waveguide without the presence of the multimode micro-racetrack resonator is measured first for normalization. The real transmission spectra of each mode in MMRR are then obtained by subtracting the reference one, as shown in Figure 9a–c. In addition, the tilt angle of the grating couplers should remain constant during the measurement. Here, the spectra are only shown in a 20 nm range for better observation of the FSRs. All three modes in the MMRR exhibit ILs below −1 dB in the measured wavelength range. Additionally, the transmission spectra of each mode contain only one set of uniformly distributed resonance peaks, suggesting that the crosstalk between modes is negligible.

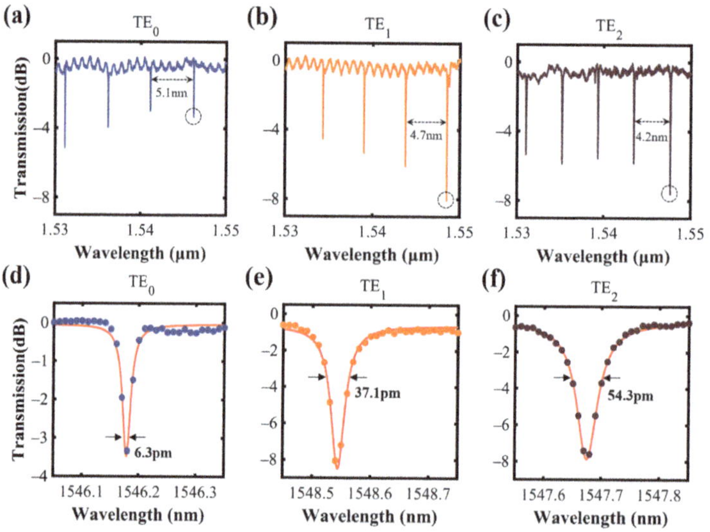

Figure 9. Transmission spectra of the fabricated MMRR. (**a**–**c**) Transmission spectra for each mode. (**d**–**f**) The measured data (scattered dots) and Lorentzian fit curves (red lines) around resonant peaks for each mode.

We select the resonant peaks near 1546.18 nm, 1548.55 nm, and 1547.67 nm for the three modes to analyze the loaded Q factors of the MMRR, and utilize Lorentzian curves to fit the measured data, as shown in Figure 9d–f. The full widths at half-maximum are 6.3 pm, 37.1 pm, and 54.3 pm for TE_0, TE_1, and TE_2 modes, corresponding to the calculated loaded Q factors of 2.3×10^5, 4.1×10^4, and 2.9×10^4, respectively. Moreover, the FSRs for the TE_0 and TE_1 modes are 5.1 nm and 4.7 nm, which are close to the calculated values. The FSR for the TE_2 mode is 4.2 nm, with a larger difference to the calculated value. This may come from the fact that the coupling length of TE_2 ADC is the smallest, so it is most susceptible to process errors. Nonetheless, the fabricated MMRR still exhibits a sufficiently large FSR in all three modes, which is highly advantageous for applications, such as wavelength division multiplexing, optical sensing, and microwave photonics.

4. Discussions

4.1. Performance Comparison

Table 2 summarizes and compares the performances of the reported MMRRs. It is evident that our MMRR exhibits the smallest effective radius. Furthermore, the fabrication of our MMRR only requires a single-step lithography and etching step. This suggests that our device is suitable for low-cost, large-scale production.

Table 2. Comparison of the performance of our device with MMRRs proposed in recent years.

Ref.	Method	Radius (μm)	FSR (nm)	Loaded Q	Modes	Fabrication	Platform
[29]	MSMMR	R_{max} = 1000, R_{min} = 30 *	1.34	\	3 TE	EBL	SOI
[30]	TO	15	3.7	5.9×10^4	3 TE	EBL	SOI
[31]	SWG	R_{max} = 600, R_{min} = 10 *	3.8	\	2 TE	\	SOI
[32]	Euler curve	R_{max} = 250, R_{min} = 130 *	0.45	4.1×10^6	1 TE	EBL	LNOI
[33]	Taper waveguide	25	2.1	7.8×10^5	2 TE	EBL	AlGaAsOI
[34]	TIR	\	6.65	7300	4 TE	EBL	SOI
This work	Cubic spline curve	8	5.1	2.3×10^5	3 TE	Lithography	SOI

* These bends are based on Euler curves with variant curvatures.

4.2. Loaded Q Factor

The loaded Q factor Q_{load} is inversely proportional to the total energy loss of the resonator and can be divided into two parts, internal loss (corresponding to Q_{int}) and coupling loss (corresponding to Q_{coup}). Therefore, the following applies:

$$\frac{1}{Q_{load}} = \frac{1}{Q_{int}} + \frac{1}{Q_{coup}} \tag{9}$$

following the calculation method shown in Figure 9, we analyze Q_{int} of the MMRR for the three modes. The resonant wavelengths are selected as 1544.87 nm, 1548.66 nm, and 1546.78 nm from the transmission spectra in Figure 7. The full widths at half-maximum are calculated as 5.9 pm, 34 pm, and 44.1 pm for TE_0, TE_1, and TE_2 modes, corresponding to Q_{int} of 2.6×10^5, 4.6×10^4, and 3.5×10^4, respectively. Equation (9) indicates two methods to increase loaded Q_{load}: (1) Decrease internal loss, which requires a high-quality SOI wafer. (2) Reduce coupling loss, which requires advanced lithography technology.

4.3. Fabrication Errors

The FSR of the TE_2 mode for the fabricated MMRR is 4.2 nm which deviates from the theoretical value (5.9 nm). It may be due to the following: (1) The field distribution of the TE_2 mode overlaps with the sidewall of the waveguide. The sidewall is relatively rough and changes the propagation performance of the TE_2 mode. (2) The parameters of the MMRR, such as waveguide widths, waveguides thicknesses, and waveguide sidewall angles, may also deviate from the designed value, which causes the error between the theoretical and experimental FSR. (3) The bus waveguide width W_{bus} is related to the group refractive index n_g and the FSR of the TE_2 mode.

5. Conclusions

In summary, we proposed and experimentally demonstrated a compact MMRR on a commercial 220 nm SOI platform, which is composed of shape-optimized MWBs and ADCs. Two cubic spline curves were chosen to define the MWB and the adjoint methods were used for optimization, achieving an effective radius of 8 μm. Three ADCs were carefully designed according to the phase-matching condition and the desired transmission performance. The fabricated MMRR exhibits high loaded Q factors of 2.9×10^4 to 2.3×10^5 and large FSRs of 4.2 to 5.1 nm for TE_0, TE_1, and TE_2 modes. The device footprint is only ~19×55 μm^2, making it highly suitable for integrated optoelectronic applications. This work demonstrates the potential of ultra-compact MMRRs for high-density integration into next-generation optical communication systems and microwave photonics. The proposed MMRR can be applied to various multimode applications, such as MDM systems, microwave photonic filters, and nonlinear optics. In addition, some improvements could be introduced in future works, such as more flexible mathematical curves, different optimization methods, and other material platforms.

Author Contributions: Conceptualization, Z.L. and C.C.; methodology, Z.L. and C.C.; software, C.C.; validation, Z.L. and C.C.; formal analysis, C.C.; investigation, Z.L.; resources, Z.L.; data curation, C.C.; writing—original draft preparation, Z.L.; writing—review and editing, Z.L. and C.C.; visualization, Z.L. and C.C.; supervision, L.Y.; project administration, X.F. and L.Y.; funding acquisition, X.F. and L.Y. All authors have read and agreed to the published version of the manuscript.

Funding: This research received no external funding.

Institutional Review Board Statement: Not applicable.

Informed Consent Statement: Not applicable.

Data Availability Statement: Relevant data are available from the authors upon reasonable request.

Conflicts of Interest: The authors declare no conflicts of interest.

Abbreviations

The following abbreviations are used in this manuscript:

MMRR	multimode micro-racetrack resonator
MWB	multimode waveguide bend
ADC	asymmetric directional coupler
FSR	free spectral range
MDM	mode division multiplexing
SWG	subwavelength grating
IL	insertion loss
TIR	total internal reflection
SOI	silicon-on-insulator
FOM	figure-of-merit
FDTD	finite-difference time-domain
ER	extinction ratio
EBL	electron beam lithography
LNOI	lithium-niobate-on-insulator
AlGaAsOI	AlGaAs-on-insulator
MSMMR	multiple single mode micro-racetracks
OSA	optical spectrum analyzer
ASE	amplified spontaneous emission
TO	transformation optics

References

1. Yao, J.; Capmany, J. Microwave Photonics. *Sci. China Inf. Sci.* **2022**, *65*, 221401. [CrossRef]
2. Marpaung, D.; Yao, J.; Capmany, J. Integrated Microwave Photonics. *Nat. Photon.* **2019**, *13*, 80–90. [CrossRef]
3. Eppenberger, M.; Messner, A.; Bitachon, B.I.; Heni, W.; Blatter, T.; Habegger, P.; Destraz, M.; De Leo, E.; Meier, N.; Del Medico, N.; et al. Resonant Plasmonic Micro-Racetrack Modulators with High Bandwidth and High Temperature Tolerance. *Nat. Photon.* **2023**, *17*, 360–367. [CrossRef]

4. Kashi, A.A.; Van Der Tol, J.J.G.M.; Lebby, M.S.; Zhang, X.; Williams, K.; Jiao, Y. Ring-Assisted Mach-Zehnder Modulator on the InP Membrane on Silicon Platform. *J. Light. Technol.* **2024**, *42*, 4553–4562. [CrossRef]
5. Green, W.M.J.; Lee, R.K.; DeRose, G.A.; Scherer, A.; Yariv, A. Hybrid InGaAsP-InP Mach-Zehnder Racetrack Resonator for Thermooptic Switching and Coupling Control. *Opt. Express* **2005**, *13*, 1651–1659. [CrossRef]
6. Xu, H.-H.; Huang, Q.-Z.; Li, Y.-T.; Yu, Y.-D.; Yu, J.-Z. Sub-Nanosecond Optical Switch Based on Silicon Racetrack Resonator. *Chin. Phys. B* **2010**, *19*, 084210. [CrossRef]
7. Tao, Z.; Tao, Y.; Jin, M.; Qin, J.; Chen, R.; Shen, B.; Wu, Y.; Shu, H.; Yu, S.; Wang, X. Highly Reconfigurable Silicon Integrated Microwave Photonic Filter towards Next-Generation Wireless Communication. *Photon. Res.* **2023**, *11*, 682. [CrossRef]
8. Yan, H.; Xie, Y.; Zhang, L.; Dai, D. Wideband-Tunable on-Chip Microwave Photonic Filter with Ultrahigh-*Q* U-Bend-Mach–Zehnder-Interferometer-Coupled Microring Resonators. *Laser Photonics Rev.* **2023**, *17*, 2300347. [CrossRef]
9. Zhao, W.; Peng, Y.; Cao, X.; Zhao, S.; Liu, R.; Wei, Y.; Liu, D.; Yi, X.; Han, S.; Wan, Y.; et al. 96-Channel on-Chip Reconfigurable Optical Add-Drop Multiplexer for Multidimensional Multiplexing Systems. *Nanophotonics* **2022**, *11*, 4299–4313. [CrossRef]
10. James, A.; Novick, A.; Rizzo, A.; Parsons, R.; Jang, K.; Hattink, M.; Bergman, K. Scaling Comb-Driven Resonator-Based DWDM Silicon Photonic Links to Multi-Tb/s in the Multi-FSR Regime. *Optica* **2023**, *10*, 832. [CrossRef]
11. Yi, X.; Zhao, W.; Li, C.; Zhang, L.; Xiang, Y.; Liu, C.; Shi, Y.; Liu, L.; Dai, D. Reconfigurable Optical Add-Drop Multiplexers for Hybrid Mode-/Wavelength-Division-Multiplexing Systems. *Adv. Photon. Nexus* **2023**, *2*, 066004. [CrossRef]
12. Mojaver, K.R.; Safaee, S.M.R.; Morrison, S.S.; Liboiron-Ladouceur, O. Recent Advancements in Mode Division Multiplexing for Communication and Computation in Silicon Photonics. *J. Light. Technol.* **2024**, *42*, 7860–7870. [CrossRef]
13. Li, C.; Liu, D.; Dai, D. Multimode Silicon Photonics. *Nanophotonics* **2019**, *8*, 227–247. [CrossRef]
14. Wu, X.; Huang, C.; Xu, K.; Shu, C.; Tsang, H.K. Mode-Division Multiplexing for Silicon Photonic Network-on-Chip. *J. Light. Technol.* **2017**, *35*, 3223–3228. [CrossRef]
15. Dai, D.; Li, C.; Wang, S.; Wu, H.; Shi, Y.; Wu, Z.; Gao, S.; Dai, T.; Yu, H.; Tsang, H. 10-Channel Mode (de)Multiplexer with Dual Polarizations. *Laser Photonics Rev.* **2018**, *12*, 1700109. [CrossRef]
16. Zhao, W.; Yi, X.; Peng, Y.; Zhang, L.; Chen, H.; Dai, D. Silicon Multimode Waveguide Crossing Based on Anisotropic Subwavelength Gratings. *Laser Photonics Rev.* **2022**, *16*, 2100623. [CrossRef]
17. Chang, W.; Zhang, M. Silicon-Based Multimode Waveguide Crossings. *J. Phys. Photonics* **2020**, *2*, 022002. [CrossRef]
18. Liu, R.; Lu, L.; Zhang, P.; Chang, W.; Liu, D.; Zhang, M. Integrated Dual-Mode 3-dB Power Splitter Based on Multimode Interference Coupler. *IEEE Photon. Technol. Lett.* **2020**, *32*, 883–886. [CrossRef]
19. Sun, S.; Fu, Y.; Li, S.; Ning, X.; Zhou, Z.; Zhang, D.; Wang, X. Ultra-Broadband Multimode Fiber-to-Chip Edge Coupler Based on Periodically Segmented Waveguides. *Opt. Lett.* **2024**, *49*, 6061. [CrossRef]
20. Zhang, R.; Li, X.; He, Y.; Chen, G.; Xiong, L.; Chen, H.; Fontaine, N.K.; Zhang, Y.; Xie, W.; Su, Y. Ultra-High Bandwidth Density and Power Efficiency Chip-To-Chip Multimode Transmission through a Rectangular Core Few-Mode Fiber. *Laser Photonics Rev.* **2023**, *17*, 2200750. [CrossRef]
21. Dai, D.; Bowers, J.E. Silicon-Based on-Chip Multiplexing Technologies and Devices for Peta-Bit Optical Interconnects. *Nanophotonics* **2014**, *3*, 283–311. [CrossRef]
22. Jiang, X.; Wu, H.; Dai, D. Low-Loss and Low-Crosstalk Multimode Waveguide Bend on Silicon. *Opt. Express* **2018**, *26*, 17680–17689. [CrossRef] [PubMed]
23. Sun, C.; Yu, Y.; Chen, G.; Zhang, X. Ultra-Compact Bent Multimode Silicon Waveguide with Ultralow Inter-Mode Crosstalk. *Opt. Lett.* **2017**, *42*, 3004. [CrossRef] [PubMed]
24. Wu, H.; Li, C.; Song, L.; Tsang, H.-K.; Bowers, J.E.; Dai, D. Ultra-Sharp Multimode Waveguide Bends with Subwavelength Gratings. *Laser Photonics Rev.* **2019**, *13*, 1800119. [CrossRef]
25. Li, W.; Chen, J.; Zhu, M.; Dai, D.; Shi, Y. Ultra-Compact Multimode Waveguide Bend With Optimized Dual Bezier Contours. *IEEE Photon. Technol. Lett.* **2023**, *35*, 1131–1134. [CrossRef]
26. Li, S.; Cai, L.; Gao, D.; Dong, J.; Hou, J.; Yang, C.; Chen, S.; Zhang, X. Compact and Broadband Multimode Waveguide Bend by Shape-Optimizing with Transformation Optics. *Photon. Res.* **2020**, *8*, 1843. [CrossRef]
27. Sun, S.; Dong, P.; Zhang, F.; Wang, J.; Zhu, N.; Shi, Y. Inverse Design of Ultra-Compact Multimode Waveguide Bends Based on the Free-Form Curves. *Laser Photonics Rev.* **2021**, *15*, 2100162. [CrossRef]
28. Sun, S.; Yang, Z.; Wang, J.; Zhang, R.; Zhang, F.; Zhu, N.; Wan, L.; Li, Z. Ultra-Sharp Silicon Multimode Waveguide Bends Based on Double Free-Form Curves. *Photonics Res. PRJ* **2022**, *10*, 1484–1490. [CrossRef]
29. Tan, J.; Xiao, H.; Ma, M.; Zhou, X.; Yuan, M.; Dubey, A.; Boes, A.; Nguyen, T.G.; Ren, G.; Su, Y.; et al. Arbitrary Access to Optical Carriers in Silicon Photonic Mode/Wavelength Hybrid Division Multiplexing Circuits. *Opt. Lett.* **2022**, *47*, 3531. [CrossRef]
30. Cai, L.; Gao, D.; Dong, J.; Hou, J.; Yang, C.; Chen, S.; Zhang, X. Compact and High Q-Factor Multimode Racetrack Ring Resonator Based on Transformation Optics. *Opt. Express* **2022**, *30*, 15766. [CrossRef]
31. Chen, Z.; Ye, M.; Zhang, L.; Zhang, Q. Compact mode-insensitive ring resonator based on SWG coupler and Euler bends. In Proceedings of the AOPC 2021: Micro-Optics and MOEMS, Beijing, China, 20–22 June 2021; Volume 12066, pp. 144–149.

32. Wei, C.; Li, J.; Jia, Q.; Li, D.; Liu, J. Ultrahigh-Q Lithium Niobate Microring Resonator with Multimode Waveguide. *Opt. Lett.* **2023**, *48*, 2465. [CrossRef] [PubMed]
33. Ye, C.; Liu, Y.; Zhou, Y.; Zhao, Y.; Zheng, Y.; Kim, C.; Oxenløwe, L.K.; Yvind, K.; Pu, M. Multimode AlGaAs-on-Insulator Microring Resonators for Nonlinear Photonics. *IEEE J. Sel. Top. Quantum Electron.* **2024**, *30*, 1–11. [CrossRef]
34. Ye, M.; Sun, C.; Yu, Y.; Ding, Y.; Zhang, X. Silicon Integrated Multi-Mode Ring Resonator. *Nanophotonics* **2021**, *10*, 1265–1272. [CrossRef]
35. Yang, S.; Jia, H.; Niu, J.; Fu, X.; Yang, L. CMOS-Compatible Ultra-Compact Silicon Multimode Waveguide Bend Based on Inverse Design Method. *Opt. Commun.* **2022**, *523*, 128733. [CrossRef]
36. Dou, T.; Yue, S.; Hou, Y.; Wang, R.; Shi, H.; Li, Y.; Feng, Y.; Zhang, Z. Ultra-Sharp Multimode Waveguide Bends Based on Superellipse Curves and Shallowly Etched Nanoslots Designed by an Intelligent Algorithm. *J. Light. Technol.* **2024**, *42*, 5249–5261. [CrossRef]
37. Liao, J.; Tian, Y.; Yang, Z.; Xu, H.; Dai, T.; Zhang, X.; Kang, Z. Broadband and Compact Silicon Multimode Waveguide Bends Based on Hybrid Shape Optimization. *J. Light. Technol.* **2024**, *42*, 4882–4891. [CrossRef]
38. Zhang, E.; Yang, S.; Zhang, L. General Waveguide Bend Design Based on Cubic Spline Interpolation and Inverse Design. *J. Light. Technol.* **2024**, *42*, 4614–4625. [CrossRef]
39. Lalau-Keraly, C.M.; Bhargava, S.; Miller, O.D.; Yablonovitch, E. Adjoint Shape Optimization Applied to Electromagnetic Design. *Opt. Express* **2013**, *21*, 21693. [CrossRef]
40. Georgieva, N.K.; Glavic, S.; Bakr, M.H.; Bandler, J.W. Feasible Adjoint Sensitivity Technique for EM Design Optimization. *IEEE Trans. Microw. Theory Tech.* **2002**, *50*, 2751–2758. [CrossRef]
41. Cheng, C.; Yang, S.; Liu, W.; Liu, S.; Huo, Y.; Fu, X.; Yang, L. Ultra-Compact Inverse Designed Multimode Waveguide Bend Based on Levelset Method. *IEEE Photonics Technol. Lett.* **2024**, *36*, 571–574. [CrossRef]
42. Teytaud, O.; Bennet, P.; Moreau, A. Discrete Global Optimization Algorithms for the Inverse Design of Silicon Photonics Devices. *Photonics Nanostruct.-Fundam. Appl.* **2022**, *52*, 101072. [CrossRef]
43. Haus, H.A.; Huang, W. Coupled-Mode Theory. *Proc. IEEE* **1991**, *79*, 1505–1518. [CrossRef]
44. Dai, D.; Bowers, J.E. Novel Ultra-Short and Ultra-Broadband Polarization Beam Splitter Based on a Bent Directional Coupler. *Opt. Express* **2011**, *19*, 18614–18620. [CrossRef]
45. McKinnon, W.R.; Xu, D.X.; Storey, C.; Post, E.; Densmore, A.; Delâge, A.; Waldron, P.; Schmid, J.H.; Janz, S. Extracting Coupling and Loss Coefficients from a Ring Resonator. *Opt. Express* **2009**, *17*, 18971. [CrossRef]

hv photonics

MDPI

Article

Compact Polarization-Insensitive 2 × 2 3 dB Quasi-Adiabatic Coupler Based on Shape Optimization

Guangchen Su [1,2], Hongliang Chen [1,2], Xin Fu [1] and Lin Yang [1,2,*]

1 Key Laboratory of Optoelectronic Materials and Devices, Institute of Semiconductors, Chinese Academy of Sciences, Beijing 100083, China; suguangchen@semi.ac.cn (G.S.); chenhongliang@semi.ac.cn (H.C.); fuxin@semi.ac.cn (X.F.)

2 Center of Materials Science and Optoelectronics Engineering, University of Chinese Academy of Sciences, Beijing 100049, China

* Correspondence: oip@semi.ac.cn

Abstract: In this study, we designed and experimentally demonstrated a compact polarization-insensitive 2 × 2 3 dB quasi-adiabatic coupler based on B-spline curves and shape optimization. By using the supermode to enable the segmented shape optimization of the coupler, we significantly reduced the computational cost of the optimization process. The numerical simulation results exhibited a power imbalance below ±0.46 dB and an insert loss (IL) of less than 0.09 dB over a broad bandwidth of 140 nm, ranging from 1490 nm to 1630 nm for both the TE and TM polarizations, with a compact coupling length of 12 μm. The experimental results showed a power splitting ratio within 3 ± 0.46 dB over the range of 1525 nm–1600 nm for the TM mode and 1576 nm–1610 nm for the TE mode. This broadband and low-loss 3 dB coupler is suitable for microwave photonic (MPW) applications, enabling efficient polarization-independent signal processing in integrated photonic systems.

Keywords: quasi-adiabatic coupler; inverse design; shape optimization; silicon photonics; waveguide devices

1. Introduction

With the rapid development of the internet, cloud computing, artificial intelligence, and other technologies, the demand for photonic integrated circuits (PICs) has increased. Silicon-on-insulator (SOI) platforms are some of the most popular platforms for actualizing low-cost PICs due to their complementary metal–oxide–semiconductor (CMOS) compatibility. Simultaneously, the high refractive index contrast between the core and cladding gives SOI platforms the ability to implement high-density integration. However, the high-refractive-index contrast and significant difference between the width and height of typical SOI waveguides have also contributed to their strong polarization dependence. In addition, 2 × 2 3 dB couplers are some of the most critical components in PICs, and they are often used in Mach–Zehnder interferometers (MZIs) [1] to split or combine light. MZIs are widely used in optical modulations [2], switches [3], optical computing [4], microwave photonics (MPW) [5,6], etc. Therefore, realizing polarization-insensitive 2 × 2 3 dB couplers is essential for realizing various polarization-insensitive devices and systems. Polarization-independent 3 dB couplers have been implemented in many ways, including subwavelength-grating (SWG)-assisted couplers [7], multimode interference (MMI) couplers [8], bent directional couplers (DCs) [9], and adiabatic couplers (ADCs) [10,11].

 https://doi.org/10.3390/photonics12030208

SWG-assisted couplers can achieve a broad bandwidth and low insertion loss (IL). Nevertheless, SWG-assisted couplers require the precise control of the shape and size of the periodic structure. MMI couplers with a broad bandwidth, small footprint, and low loss are only utilized as 1 × 2 power splitters. Bent DCs, which consist of a cascade of bent waveguides, also require a high fraction precision. ADCs operate based on the evolution of individual modes, resulting in a large bandwidth and high fabrication tolerance. However, to avoid the excitation of undesired modes, ADCs often have very long lengths. Therefore, the realization of broadband, low-loss, compact, and highly manufacturing-tolerant 2 × 2 3 dB couplers remain to be further explored.

Recently, inverse design has emerged as a powerful approach in photonics design for obtaining compact and high-performance devices [12], such as grating couplers [13], wavelength (de)multiplexers [14], and mode (de)multiplexers [15]. Shape optimization based on the adjoint method [16] is a gradient-based algorithm, and the shape of the device is determined by a designer-specified curve. So far, there have been limited investigations into the utilization of adjoint shape optimization for the design of polarization-insensitive 3 dB couplers. B-spline curves [17] have some advantages over Bézier curves, which have been used in shape optimization for many photonic devices [18,19]. Compared with Bézier curves, B-spline curves have more flexibility when the number of control points is very large, since the degree of the curve function is not necessarily equal to the number of control points minus one and a single control point no longer affects the shape of the whole curve. However, more control points mean more time is needed to compute the gradient. In addition, for B-spline curves, the number of control points, the order of the curves, and the distribution of the knots can all be freely chosen, meaning that more manual parameter adjustments are required. Therefore, in this study, we combine B-splines and adjoint shape optimization to calculate a large number of parameters quickly while utilizing the advantages of B-spline curves.

In this work, we propose and experimentally demonstrate a compact polarization-insensitive 2 × 2 3 dB quasi-adiabatic coupler using shape optimization based on the adjoint method and B-spline curves. During the optimization, we introduced the supermode of the two-waveguide system as the forward and adjoint sources to reduce the number of simulations needed for this multi-objective problem. In addition, we separately optimized the coupling region and the input region of the coupler. Our coupler has a mode coupling length of 12 μm and a minimum feature size of 150 nm. The numerical simulation resulted in an insert loss (IL) of less than 0.09 dB and a splitting ratio between 45% and 55% over a broad bandwidth of 140 nm for both the TE and TM polarizations. We fabricated the devices on the SOI platform. The experimental results show that the TM mode achieved a power splitting ratio within 3 ± 0.46 dB (50% ± 5%) over the wavelength range of 1525 nm–1600 nm, and the TE mode achieved the same power splitting ratio from 1576 nm to 1610 nm.

2. Design and Analysis

The schematic for our 2 × 2 3 dB coupler is shown in Figure 1. The coupler is designed based on a 220 nm thick SOI platform with a 3 μm thick buried oxide layer and a 3 μm thick upper silica cladding layer. Figure 1a,b show a three-dimensional view and a cross-sectional view of the coupler, respectively. Figure 1c shows a top view of the coupler, showing that the device consists of three regions. In region 1, one straight waveguide and one S-shaped waveguide constitute the input section of the device. The widths of these two waveguides are W_1 = 350 nm and W_2 = 550 nm, and the gap between them gradually decreases from g1 = 1.65 μm to g2 = 150 nm over the length of L_1 = 30 μm. To make the device more compact, we optimized the shape of the standard S-shaped waveguide by two B-spline curves $C_{in\,i}(i = 1,2)$. After optimization, the length of the input section was reduced to 15 μm. In region 2, a two-waveguide system constitutes the coupling section of the device and is also the most critical region of the design. In our design, the shape of these two waveguides is defined by four B-spline curves $C_i(i = 1,2,3,4)$, and the control points P_{ij} of these curves are determined by shape-based optimization. The length of this section is L_2 = 12 um. In region 3, two symmetrical S-shaped waveguides with width W_3 = 450 nm and length L_3 = 10 μm separate the gap from g_2 = 150 nm to g_3 = 1.6 μm, forming the output section of the device.

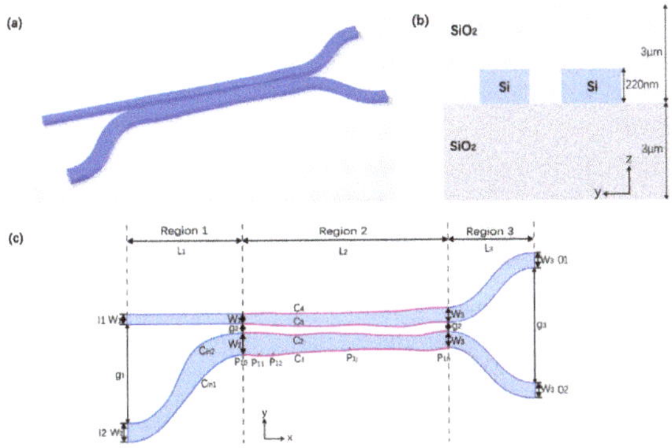

Figure 1. (**a**) Three-dimensional schematic, (**b**) cross-sectional view, and (**c**) top view of the 2 × 2 3 dB coupler.

Figure 2 illustrates the mode profiles for each section's input and output ports. In this paper, we order the modes of the two-waveguide system according to the effective index from largest to smallest. When TE_0-/TM_0-mode light is launched into input I1 on the narrow waveguide, the second-/fourth-order supermode of the two asymmetrical waveguides is initially excited in the coupling section and then evolves into the odd supermode of the two symmetrical waveguides. At the two outputs O1 and O2, the light is split evenly and has a π phase difference. Similarly, when TE_0-/TM_0-mode light is launched into input I2 on the wide waveguide, the first-/third order supermode of the two asymmetrical waveguides is excited and then evolves into the even supermode of the two symmetrical waveguides. At the two outputs O1 and O2, the light has the same intensity and phase.

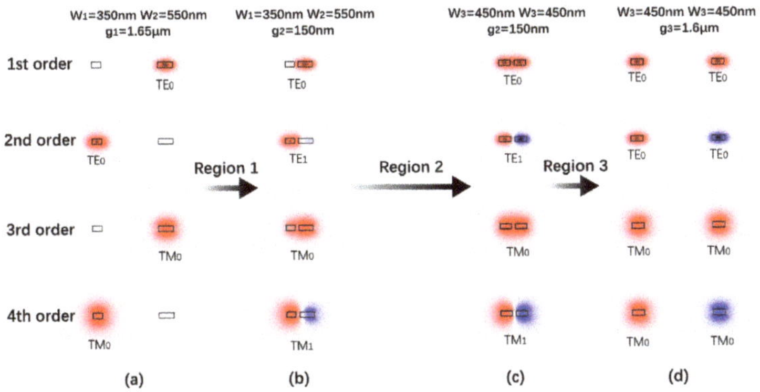

Figure 2. The mode profiles of each section's input and output ports: (**a**) input port of region 1; (**b**) input port of region 2; (**c**) output port of region 2; (**d**) output port of region 3.

ADCs usually require a long coupling region to avoid the input supermode coupling into the other unexpected supermode with the same polarization during its evolution toward the output port. We use an inverse design to shorten the evolution length while maintaining the adiabatic evolution of the excited supermodes along the coupling region. The expression of the four B-spline curves is:

$$C_i(t) = \sum_{j=0}^{16} P_{i,j} N_{j,3}(t),\ t \in [0,1] \tag{1}$$

where P_{ij} represents the (j+1)-th control point of curve C_i, and the coordinates of P_{ij} are (x_{ij}, y_{ij}). $N_{j,3}(t)(j = 0, 1, \ldots, 16)$ are the third-order B-spline basis functions. There are several ways to define B-spline basis functions, and in this work, we use the Cox–de Boor recursion formula [17] to define them:

$$\begin{aligned} N_{i,1}(t) &= \begin{cases} 1 & t_i < x < t_{i+1} \\ 0 & Otherwise \end{cases} \\ N_{i,k}(t) &= \frac{t-t_i}{t_{i+k-1}-t_i} N_{i,k-1}(t) + \frac{t_{i+k}-t}{t_{i+k}-t_{i+1}} N_{i+1,k-1}(t) \end{aligned} \tag{2}$$

where $N_{i,k}(i = 0, 1, \ldots, n)$ are the i-th B-spline basis functions of order k and also are piecewise polynomials of degree k-1. These piecewise polynomials are joined by the knot vector, which is a sequence of non-decreasing real numbers $\vec{T} = (t_0, t_1, \ldots, t_m)$, and $m = n + k + 1$.

There are three key restrictions on the parameters of the four curves in Equation (1): (1) To guarantee a smooth connection between the coupling waveguides and the input/output waveguides, the first control points and the last control points are fixed, and the endpoints of the curves coincide with these control points. (2) More control points and a lower order will introduce more details to the curve, but they may also introduce more difficulties in terms of fabrication. A growth in the number of control points will also improve the time required to calculate the gradient. On the other hand, fewer control points and a higher order will result in a smoother curve but will reduce the degrees of freedom in the parameter space. Consequently, the number of control points and the order of the curve were appropriately chosen to be 17 and 3 for balancing the flexibility and smoothness of the curve. (3) If the control points of curves C_2 and C_3 are allowed to move freely, the gap between them may not meet the fabrication constraints. Therefore, we set the gap between C_3 and C_2 constant at 150 nm, which means $P_{3,j} = P_{2,j} + (0, g_2)$. Finally, the number of control points that are available for adjustment is 45, and both the x and y coordinates of these points will be optimized.

The figure of merit (FOM) for the coupling region is defined as follows:

$$FOM = \sum_{i=1}^{4} FOM_{coupling\ i} \tag{3}$$

where $FOM_{coupling\ i}$ is defined as the transmission onto the i-th supermode of the two symmetrical waveguides, as illustrated in Figure 2c, where the input is the i-th supermode of the two asymmetrical waveguides, as illustrated in Figure 2b. Therefore, the target value of $FOM_{coupling\ i}$ in this optimization was set to 1. To improve the device's performance over a wide bandwidth, optimization was carried out at five wavelengths $\lambda_j(j = 1, 2, 3, 4, 5)$, i.e., 1.5 µm, 1.525 µm, 1.55 µm, 1.575 µm, and 1.6 µm. Moreover, we used the root mean square error to calculate the error between the true value and the target value for this broadband optimization. The definition of $FOM_{coupling\ i}$ is as follows:

$$FOM_{coupling\ i} = 1 - \left(\frac{1}{5}\sum_{j=1}^{5}\left(1 - T_i(\lambda_j)\right)^2\right)^{\frac{1}{2}} \quad (4)$$

The three-dimensional finite-difference time-domain (3D-FDTD) method and adjoint method are applied to update the gradient of the FOM. During each iteration, the gradient of the FOM with respect to the permittivity for each mode is calculated by two electric field distributions, which are obtained from two simulations consisting of one forward simulation and one adjoint simulation. Consequently, a total of eight simulations are needed for the four modes. Next, the gradient of the FOM with respect to the design parameters is calculated by the chain rule, and here, the design parameters refer to the coordinates of the control points. After obtaining the gradient, the L-BFGS-B algorithm is employed to update the design parameters.

Before starting the optimization, the shape of the four curves in the coupling region was initially chosen to be straight, which also means that the two waveguides were linearly tapered. To realize this shape, the initial coordinates of the control points were uniformly distributed between the first and the last points. The final optimized curve shape is shown in Figure 3a. Figure 3b illustrates the evolution of the FOMs with iterations for four modes. After 135 iterations, the FOMs for modes 1 to 4 were 0.9955, 0.9911, 0.9945, and 0.9860, respectively. The iteration procedure took 44 h on a 2.10 Ghz Intel Core 17-12700 PC with 32 GB RAM.

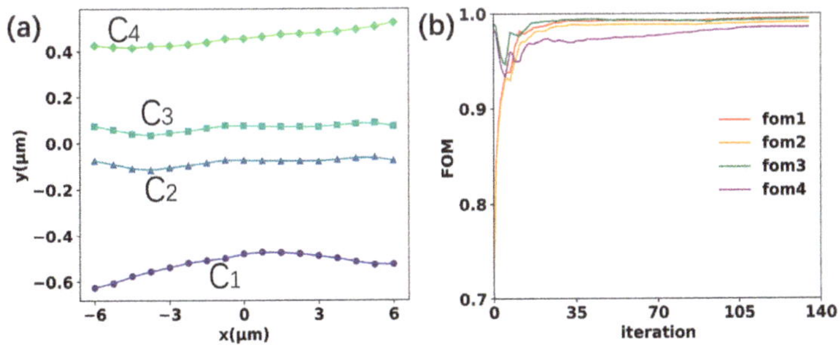

Figure 3. (**a**) Curve shape of the optimized coupling region, (**b**) four modes' FOM as a function of iterations.

Figure 4a,b show the simulated results of the 2 × 2 3 dB coupler with a standard S-shaped input waveguide of 30 μm length before and after optimizing the shape of the coupling waveguides. As shown in Figure 4a, the coupling region consisting of two linear tapered waveguides with a length of 12 μm was too short for 3 dB power splitting. After optimizing the shape of the coupling waveguides, the power splitting ratio was maintained within 50% ± 5% for both the TE and TM modes over a wavelength range of 1490 nm–1625 nm, with the IL being lower than 0.06 dB.

When two waveguides get closer, a longer length helps to achieve a lower insertion loss (IL) by reducing bending losses or the excitation of undesired modes. To further minimize the footprint of the device while maintaining low loss, we optimized the S-shaped waveguide in the input region using the same algorithm as the coupling region. The number of control points and the order of the curves were chosen to be 11 and 3, respectively. The optimization was also performed at five equally spaced wavelengths from 1500 nm to 1600 nm. The shapes of the input waveguides of the final optimized curves are shown in Figure 5a. After 20 iterations, the FOMs for modes 1 to 4 were 0.9996, 0.9985, 0.9952, and 0.9929, respectively. The iteration curves of the FOM are shown in Figure 5b. The iteration procedure took 29 h on a dual 2.90 Ghz Intel Xeon Platinum 8268 server with 256 GB RAM.

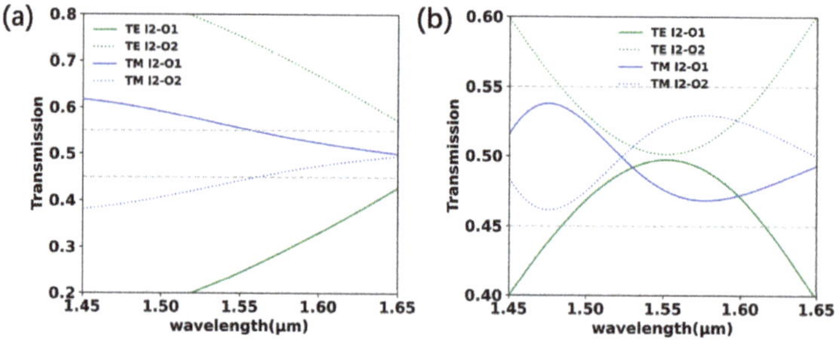

Figure 4. Simulated power splitting ratios of the coupler as a function of wavelengths with a standard S-shaped input waveguide and (**a**) a coupling region in the initial shape; (**b**) a shape-optimized coupling region. The horizontal gray dashed lines for reference are located at 0.45 and 0.55, the same below.

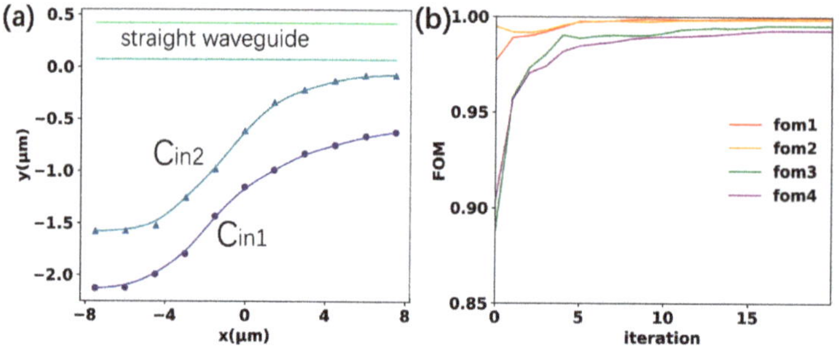

Figure 5. (**a**) Shape of the curves of the optimized input region, (**b**) FOM of the four modes as a function of iterations.

Figure 6a shows the simulated results of the 2 × 2 3 dB coupler with a standard S-shaped input waveguide with a 15 μm length and a shape-optimized coupling region. Shortening the length of the input region significantly reduced the performance of the TM mode. Figure 6b shows the simulated results of the 2 × 2 3 dB coupler with both the input region and coupling region optimized. After further optimizing the shape of the input waveguides, the power splitting ratio was maintained within 50% ± 5% for both modes over a wavelength range of 1490 nm–1630 nm, with a maximum IL of 0.09 dB. Figure 7a,b show the power splitting ratio of the final optimized device with a waveguide width variation of $\Delta w = 0, \pm 10$, and ± 20 nm when light was launched into input I2. The results show that the device can tolerate fabrication errors up to ±10 nm in terms of width variation. The results for a ±10 nm variation in waveguide thickness are displayed in Figure 8. It can be seen that the spectra of the TM mode were more affected by the thickness of the waveguide.

Figure 9a demonstrates the simulated light propagation profiles of the final optimized device for both polarizations at a 1550 nm wavelength when the light was launched into input I1 and I2, respectively. It can be seen that the operation of the TE mode was based on adiabatic evolution. In the case of the TM mode, the 3 dB power splitting was actually realized based on mode coupling in a quasi-adiabatic regime.

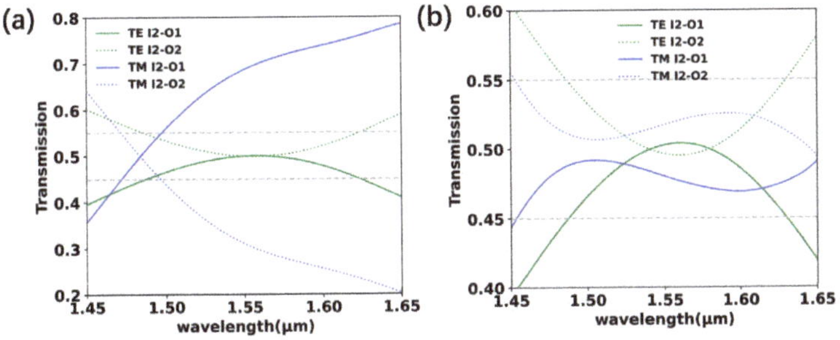

Figure 6. Simulated power splitting ratio of the coupler as a function of wavelengths with a shape-optimized coupling region and (**a**) a standard S-shaped input waveguide; (**b**) a shape-optimized input region.

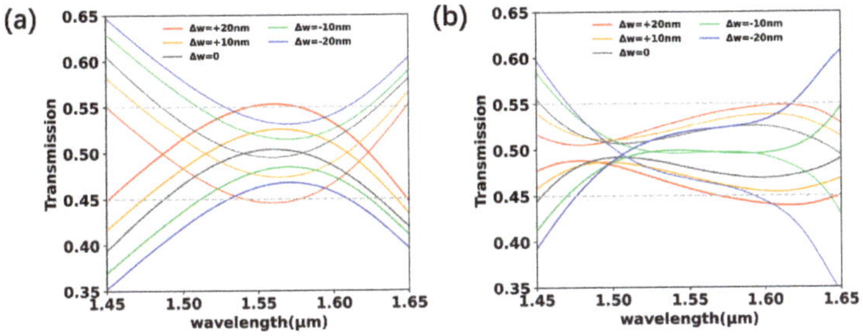

Figure 7. The power splitting ratio with a waveguide width variation of Δw = 0, ±10, and ±20 nm: (**a**) TE mode; (**b**) TM mode.

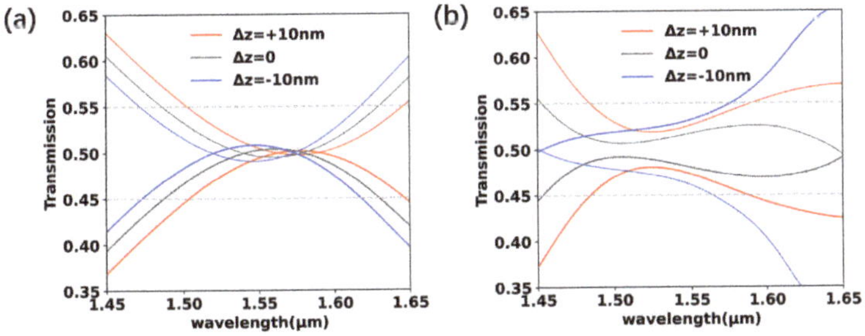

Figure 8. The power splitting ratio with a waveguide thickness variation of Δz = 0, ±10 nm: (**a**) TE mode; (**b**) TM mode.

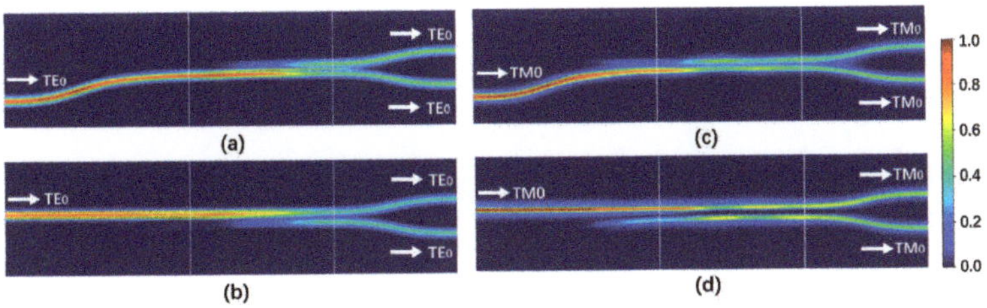

Figure 9. FDTD-simulated light propagation profiles at 1550 nm wavelength. Input 2: (**a**) TE mode; (**b**) TM mode. Input 1: (**c**) TE mode; (**d**) TM mode.

3. Fabrication and Experimental Results

The device was fabricated through the SOI platform with a top silicon thickness of 220 nm and a buried oxide layer thickness of 3 μm. The fabrication process began with electron beam lithography (EBL) to define the pattern, followed by inductively coupled plasma (ICP) etching to transfer the pattern onto the silicon layer. Finally, a 3 μm thick upper silica cladding layer was deposited by plasma-enhanced chemical vapor deposition (PECVD). Figure 10a shows a scanning electron microscope (SEM) image of a fabricated 2 × 2 3 dB coupler. Figure 10b shows a microscope image of an unbalanced Mach–Zehnder interferometer (MZI), which was used to characterize the power splitting ratios of the device, and the difference in the length between the two arms was ΔL = 150 μm for the TE mode and ΔL = 250 μm for the TM mode.

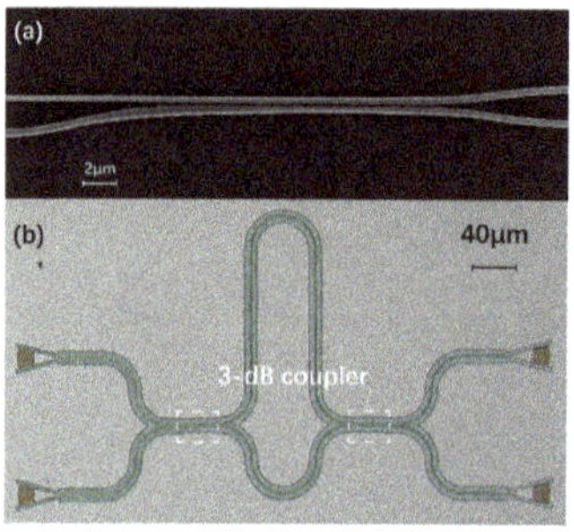

Figure 10. (**a**) SEM image of the fabricated coupler. (**b**) Optical microscope photos of an unbalanced MZI.

To characterize the performance of the fabricated devices, an amplified spontaneous emission (ASE) broadband light source with an operating wavelength range of 1525 nm–1610 nm and an optical spectrum analyzer (OSA) was employed. The measured spectra of the unbalanced MZI and the power splitting ratios K of the 3 dB coupler are displayed in Figure 11. Here, K was calculated from the extinction ratios (ERs) of the measured spectra according to the following equation [9]:

$$K = \frac{1}{2} \pm \frac{1}{2}\sqrt{\frac{1}{10^{\frac{ER}{10}}}} \tag{5}$$

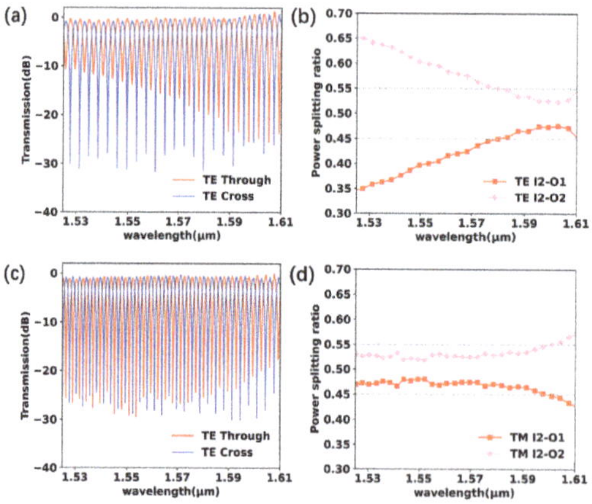

Figure 11. Measured MZI transmission spectra: (**a**) TE mode; (**c**) TM mode; power splitting ratios calculated from the ERs: (**b**) TE mode; (**d**) TM mode.

As displayed in Figure 11, the TM mode achieved a power splitting ratio within 50% ± 5%, ranging from 1525 nm to 1600 nm. Meanwhile, the TE mode only achieved the desired power splitting ratio in the range of 1576 nm–1610 nm. The measured ILs of a single device for both polarizations were less than 1 dB. The results suggest that the boundary of the fabricated 2 × 2 3 dB couplers deviated from the design. Comparison with the fabrication tolerance analysis in Figures 7 and 8 does not provide a good explanation for the apparent deviation in the experimental results of the TE mode from the simulation results. We believe that the deviation came from a particular kind of fabrication imperfection in the coupling region. We modeled the deviated curves $C_{error\ i}(i = 1,2,3,4)$ by adding a variable controlled by the parameter η on the curves $C_i(i = 1,2,3,4)$. Specifically, the curve equation can be defined as follows:

$$\begin{aligned} C_{error\ i}(t) &= \sum_{j=0}^{16} P_{error\ i,j} N_{j,3}(t),\ t \in [0,1] \\ P_{error\ i,j}(t) &= \eta \times P_{i,j} + (1-\eta) \times \left[P_{i,0} + \frac{i}{16}(P_{i,17} - P_{i,0}) \right] \end{aligned} \quad (6)$$

where η was used to control the generation of manufacturing error, and η ranged from 0.8 to 1.2. When η was less than 1, the curve approached towards the line segment whose endpoints coincided with the curve's endpoints, appearing as under-etching with a non-uniform magnitude at different locations. Conversely, when η was greater than 1, the curve moved away from the line segment, appearing as over-etching with a non-uniform magnitude at different locations. Figure 12 illustrates the deviated curves and transmission spectra for different η values. As is apparent in Figure 12b,c, the power splitting ratio in the TE mode was more sensitive to η than that in the TM mode. In addition, the experimental results of the TE mode were consistent with η < 1, indicating that the error came from the curve shape deviation caused by localized under-etching.

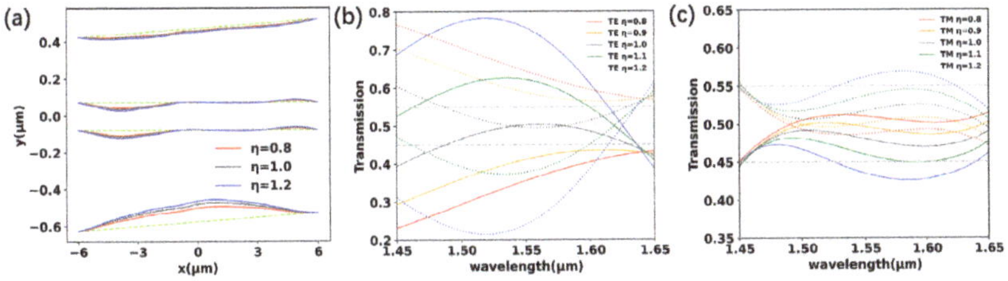

Figure 12. Illustration of (**a**) the shape, (**b**) TE-mode transmission spectra, and (**c**) TM-mode transmission spectra of the coupler for different values of η.

From a design perspective, we considered the following methods to mitigate the effects of such deviations: (1) appropriately increase the length of the coupling region; (2) not optimize the shape of the gap (curves C_2 and C_3) to reduce the effect of the simultaneous deviation of curves on both sides of the waveguide; (3) change the number of control points and the order to further improve the smoothness of the curves; (4) add some random errors during the iteration process to improve the robustness of the optimization results.

Table 1 summarizes the performance of our device and state-of-the-art polarization-insensitive 3 dB couplers built on the SOI platform with a 220 nm core silicon thickness. From the simulation results, our optimized design resulted in a more compact 2 × 2 3 dB coupler with broader bandwidth and lower loss. However, the fabrication tolerance of our design was not as high as expected, making the experimental results of the TE mode deviate excessively from the design.

Table 1. Comparison of the performance of our device with polarization-insensitive 3 dB couplers proposed in recent years.

Ref.	Structure	Type	Footprint (μm^2)	IL (dB) [1]		Bandwidth (nm)		Fabrication Tolerance (nm)
				Sim.	Exp.	Sim.	Exp.	
[7]	SWG-assisted coupler	2 × 2	35.9 × 3	<1	NA	1260–1420 (3 ± 0.5)	NA	±15
[8]	MMI coupler	1 × 2	1.8 × 1.3 (18 × 4)	0.11	0.26	1300–1700 (3 ± 0.33)	1500–1580 (3 ± 0.23)	±20
[18]	MMI coupler	1 × 2	10 × 2.5	0.6	0.63	1260–1675 (NA)	1260–1675 (3 ± 0.37)	±40
[9]	Bent DC	2 × 2	<50 μm	NA	<1	1520–1630 (3 ± 0.46)	1520–1600 (3 ± 1)	±10
[10]	ADC	2 × 2	125 × 3.4	0.17	<1	1500–1600 (3 ± 0.4)	1500–1600 (3 ± 0.6)	NA
[11]	ADC	2 × 2	53.7 × 3.3	NA	NA	NA	1490–1565 (3 ± 0.5)	NA
This work	Inverse design (2 × 2)	2 × 2	37 × 3.4	0.09	<1	1490–1630 (3 ± 0.46)	TE: 1576–1610(3 ± 0.46) TM: 1525–1600(3 ± 0.46)	±10

[1] NA: not available; Sim: simulation results; Exp: experiment results.

4. Conclusions

In conclusion, we designed and experimentally demonstrated a compact polarization-insensitive 2 × 2 3 dB quasi-adiabatic coupler based on B-spline curves and shape-based optimization. The coupling length of the device is only 12 μm, and the total length is 37 μm. By applying supermode and segmented shape optimization, we significantly reduced the computational cost of the optimization process. The numerical simulation results indicate the maximum power imbalance was 1.2 dB, with negligible ILs over a broad bandwidth of 140 nm, ranging from 1490 nm to 1630 nm for both the TE mode and TM mode. The measurement results revealed that the errors caused by fabrication imperfections were more significant than expected. Therefore, we plan to enhance the device's robustness against fabrication deviations in a further optimization process.

Author Contributions: Conceptualization, G.S. and H.C.; methodology, G.S.; software, G.S.; validation, G.S.; formal analysis, G.S.; investigation, G.S.; resources, L.Y.; data curation, G.S.; writing—original draft preparation, G.S.; writing—review and editing, H.C. and X.F.; visualization, G.S.; supervision, L.Y.; project administration, X.F. and L.Y.; funding acquisition, X.F. and L.Y. All authors have read and agreed to the published version of the manuscript.

Funding: This research was funded by the Strategic Priority Research Program of the Chinese Academy of Sciences under grant XDB43000000.

Institutional Review Board Statement: Not applicable.

Informed Consent Statement: Not applicable.

Data Availability Statement: Relevant data are available from the authors upon reasonable request.

Conflicts of Interest: The authors declare no conflicts of interest.

References

1. Dainesi, P.; Kung, A.; Chabloz, M.; Lagos, A.; Fluckiger, P.; Ionescu, A.; Fazan, P.; Declerq, M.; Renaud, P.; Robert, P. CMOS compatible fully integrated Mach-Zehnder interferometer in SOI technology. *IEEE Photonics Technol. Lett.* **2000**, *12*, 660–662. [CrossRef]

2. He, M.; Xu, M.; Ren, Y.; Jian, J.; Ruan, Z.; Xu, Y.; Gao, S.; Sun, S.; Wen, X.; Zhou, L.; et al. High-performance hybrid silicon and lithium niobate Mach–Zehnder modulators for 100 Gbit s^{-1} and beyond. *Nat. Photonics* **2019**, *13*, 359–364. [CrossRef]
3. Jia, H.; Yang, S.; Zhou, T.; Shao, S.; Fu, X.; Zhang, L.; Yang, L. WDM-compatible multimode optical switching system-on-chip. *Nanophotonics* **2019**, *8*, 889–898. [CrossRef]
4. Shen, Y.; Harris, N.C.; Skirlo, S.A.; Prabhu, M.; Baehr-Jones, T.; Hochberg, M.; Sun, X.; Zhao, S.; Larochelle, H.; Englund, D.R.; et al. Deep learning with coherent nanophotonic circuits. *Nat. Photonics* **2017**, *11*, 441–446. [CrossRef]
5. Marpaung, D.; Yao, J.; Capmany, J. Integrated microwave photonics. *Nat. Photonics* **2019**, *13*, 80–90. [CrossRef]
6. Liu, Y.; Choudhary, A.; Marpaung, D.A.; Eggleton, B.J. Integrated microwave photonic filters. *Adv. Opt. Photonics* **2020**, *12*, 485–555. [CrossRef]
7. Pérez-Armenta, C.; Ortega-Moñux, A.; Luque-González, J.M.; Halir, R.; Reyes-Iglesias, P.J.; Schmid, J.H.; Cheben, P.; Molina-Fernández, Í.; Wangüemert-Pérez, J.G. Polarization-independent multimode interference coupler with anisotropy-engineered bricked metamaterial. *Photonics Res.* **2022**, *10*, A57–A65. [CrossRef]
8. Nair, D.P.; Ménard, M. A Compact Low-Loss Broadband Polarization Independent Silicon 50/50 Splitter. *IEEE Photonics J.* **2021**, *13*, 1–7. [CrossRef]
9. Chen, X.; Liu, W.; Zhang, Y.; Shi, Y. Polarization-insensitive broadband 2 × 2 3 dB power splitter based on silicon-bent directional couplers. *Opt. Lett.* **2017**, *42*, 3738–3740. [CrossRef] [PubMed]
10. Wang, Y.; Xu, L.; Yun, H.; Ma, M.; Kumar, A.; El-Fiky, E.A.; Li, R.; Abadíacalvo, N.; Chrostowski, L.; Jaeger, N.A.; et al. Polarization-Independent Mode-Evolution-Based Coupler for the Silicon-on-Insulator Platform. *IEEE Photonics J.* **2018**, *10*, 1–10. [CrossRef]
11. Chung, H.C.; Chen, C.H.; Hung, Y.J.; Tseng, S.Y. Compact polarization-independent quasi-adiabatic 2 × 2 3 dB coupler on silicon. *Opt. Express* **2022**, *30*, 995–1002. [CrossRef] [PubMed]
12. Li, Z.; Zhou, Z.; Qiu, C.; Chen, Y.; Liang, B.; Wang, Y.; Liang, L.; Lei, Y.; Song, Y.; Jia, P.; et al. The Intelligent Design of Silicon Photonic Devices. *Adv. Opt. Mater.* **2024**, *12*, 2301337. [CrossRef]
13. Su, L.; Trivedi, R.; Sapra, N.V.; Piggott, A.Y.; Vercruysse, D.; Vučković, J. Fully-automated optimization of grating couplers. *Opt. Express* **2018**, *26*, 4023–4034. [CrossRef] [PubMed]
14. Piggott, A.; Lu, J.; Lagoudakis, K.; Petykiewicz, J.; Babinec, T.; Vuckovic, J. Inverse design and demonstration of a compact and broadband on-chip wavelength demultiplexer. *Nat. Photonics* **2015**, *9*, 374–377. [CrossRef]
15. Yang, K.Y.; Shirpurkar, C.; White, A.D.; Zang, J.; Chang, L.; Ashtiani, F.; Guidry, M.A.; Lukin, D.M.; Pericherla, S.V.; Yang, J.; et al. Multi-dimensional data transmission using inverse-designed silicon photonics and microcombs. *Nat. Commun.* **2022**, *13*, 7862. [CrossRef] [PubMed]
16. Lalau-Keraly, C.M.; Bhargava, S.; Miller, O.D.; Yablonovitch, E. Adjoint shape optimization applied to electromagnetic design. *Opt. Express* **2013**, *21*, 21693–21701. [CrossRef] [PubMed]
17. Lyche, T.; Manni, C.; Speleers, H. Foundations of Spline Theory: B-Splines, Spline Approximation, and Hierarchical Refinement. In *Splines and PDEs: From Approximation Theory to Numerical Linear Algebra*, 1st ed.; Lyche, T., Manni, C., Speleers, H., Eds.; Springer: Cham, Switzerland, 2018; Volume 2219, pp. 1–76.
18. Zhao, S.; Han, S.; Zhu, M.; Chen, J.; Shi, Y. Ultra-broadband dual-polarization and arbitrary ratio power splitters based on Bezier curve optimized multimode interference. *Opt. Lett.* **2023**, *48*, 1331–1334. [CrossRef] [PubMed]
19. Li, W.; Chen, J.; Zhu, M.; Dai, D.; Shi, Y. Ultra-Compact Multimode Waveguide Bend With Optimized Dual Bezier Contours. *IEEE Photonics Technol. Lett.* **2023**, *35*, 1131–1134. [CrossRef]

photonics

MDPI

Article

Silicon Nitride Spot-Size Converter with Coupling Loss < 1.5 dB for Both Polarizations at 1W Optical Input

Enge Zhang [1], Yu Zhang [2], Lei Zhang [1] and Xu Yang [2,*]

[1] State Key Laboratory of Information Photonics and Optical Communications, School of Integrated Circuits, Beijing University of Posts and Telecommunications, Beijing 100876, China; zhangenge@bupt.edu.cn (E.Z.); zhang-lei@bupt.edu.cn (L.Z.)

[2] The 54th Research Institute of China Electronics Technology Group Corporation, Shijiazhuang 050081, China; zyzhangyu1996@163.com

* Correspondence: yangxumail@foxmail.com

Abstract: Microwave photonics (MWP) applications often require a high optical input power (>100 mW) to achieve an optimal signal-to-noise ratio (SNR). However, conventional silicon spot-size converters (SSCs) are susceptible to high optical power due to the two-photon absorption (TPA) effect. To overcome this, we introduce a silicon nitride (SiN) SSC fabricated on a silicon-on-insulator (SOI) substrate. When coupled to a tapered fiber with a 4.5 μm mode field diameter (MFD), the device exhibits low coupling losses of <0.9 dB for TE modes and <1.4 dB for TM modes at relatively low optical input power. Even at a 1W input power, the additional loss is minimal, at approximately 0.1 dB. The versatility of the SSC is further demonstrated by its ability to efficiently couple to fibers with MFDs of 2.5 μm and 6.5 μm, maintaining coupling losses below 1.5 dB for both polarizations over the entire C-band. This adaptability to different mode diameters makes the SiN SSC a promising candidate for future electro-optic chiplets that integrate heterogeneous materials such as III-V for gain and lithium niobate for modulation with the SiN-on-SOI for all other functions using advanced packaging techniques.

Keywords: silicon photonics; spot-size converter; microwave photonics

1. Introduction

Driven by the demand for compact, low-cost, and high-speed systems, silicon photonics integrated circuits (PICs) have emerged as a promising solution due to their compactness and compatibility with complementary metal oxide semiconductor (CMOS) fabrication processes. (All abbreviations used in the manuscript are listed in Table A1.) Despite offering a versatile toolkit for building complex systems [1–6], silicon photonics suffers from three main limitations. Firstly, it needs external material to build the laser source. Secondly, the 3 dB bandwidth of silicon modulators is limited and the insertion loss and half-wave voltage are relatively high. Thirdly, the propagation loss is relatively high (~1 dB/cm), and it cannot handle high-power input due to the two-photon absorption (TPA) effect. To address these limitations, hybrid PICs have emerged as a compelling solution. Figure 1 illustrates a typical electro-optic chiplet for microwave photonics (MWP) applications. A high-power laser, generated by a III-V semiconductor gain medium and a silicon nitride (SiN) resonant cavity [7], is coupled to a thin-film lithium niobate (TFLN) modulator for high-speed modulation with a low half-wave voltage and high linearity. The modulated light is subsequently directed to a silicon-on-insulator (SOI) chip with a SiN layer for further optical processing [8,9].

Figure 1. Schematic of typical electro-optic chiplet for MWP applications.

To optically connect chips, efficient coupling structures like grating couplers (GCs) and spot-size converters (SSCs) are essential [10–26]. GCs offer versatility with flexible placement and relaxed alignment tolerances, making them ideal for wafer-level testing and packaging [10,16]. However, their coupling efficiency is limited and sensitive to wavelength and polarization [19]. In contrast, SSCs provide high coupling efficiency, a wide operating range, and low polarization sensitivity, making them a popular choice [19,21].

To effectively couple light between chips with small MFDs, such as those found in TFLN and III-V gain media (typically 2.5–6.5 μm in size) [27–30], SSCs are indispensable. While SiN-based SSCs offer certain advantages like immunity to the TPA effect and easier fiber mode matching due to their lower refractive index contrast, this low contrast also limits the potential for compact designs [31]. The enhancement of edge coupling performance on silicon nitride platforms has been extensively investigated [19,32–36]. T. Zhu et al. achieved a low-loss SSC using an inverted taper structure [32], but this design resulted in a relatively long device length of 500 μm. However, its broadband and low-loss characteristics provided an important reference for subsequent research. To further optimize the performance of the SSC, J. Fernández et al. employed an inverted lateral taper structure, reducing the SSC length to 150 μm [33]. B. Bhandari et al. achieved a low-loss coupling efficiency with a device length of only 76 μm using a multi-taper structure [19]. These studies demonstrate that an appropriate structural design can effectively reduce both device length and insertion loss. Furthermore, researchers have explored the optimization of vertical structures. For example, Y. Li et al. combined silicon nitride with silicon oxynitride (SiON) inverted-taper structures to improve coupling efficiency [34]. Y. Liang utilized an air trench to achieve broadband and low-loss edge couplers [35]. N. Kohli et al. employed a stack of SiN/SiO_2 alternating layers to attain remarkable efficiency and broadband-coupling capabilities [36]. H. Huang et al. leveraged a 3D-nanoprinted interposer to realize a photonic chiplet interconnection system [37]. While these methods have been shown to be effective, some of them present significant challenges in terms of manufacturing feasibility and practical application. Therefore, developing simple, efficient, and process-friendly methods to realize compact, efficient, and polarization-insensitive edge couplers remains an important goal.

This paper proposes a simple and effective design approach for compact SiN SSCs incorporating multiple linear taper segments. A 123 μm long SiN SSC is designed. Its 250 nm wide tip is deliberately extended by 100 microns to improve tolerance during the dicing and polishing processes. The designed SSC exhibits low polarization-dependent loss (PDL), with TE and TM coupling losses of 0.9 dB and 1.4 dB, respectively, when coupled to a 4.5 μm MFD fiber. In particular, the designed SSC demonstrates robustness to high power, maintaining low-loss at 1W. It also accommodates a wide range of fiber MFDs, exhibiting coupling losses of less than 1.5 dB for fibers with MFDs ranging from 2.5 μm to 6.5 μm. The SSC has 3 dB alignment tolerances of 3.6 μm transversely and 3.4 μm longitudinally. This simple and compact SSC structure is useful for the design of electro-optic chiplets requiring high-power optical input.

2. Design Principle

The structure of the SSC is shown in Figure 2. It is divided into three sections. The first section consists of a 100 μm long narrow tip waveguide (NTW) with the same width along the waveguide, which has an MFD similar to that of the access fiber. The second part is the mode conversion waveguide (MCW), which consists of multiple segments of linearly tapered waveguides. The third part is a single-mode SiN waveguide with a width of ~1 μm. For a given fiber MFD, the main design task for the SiN SSC is to select an optimal tip width for the NTW and to design a compact and low-loss MCW.

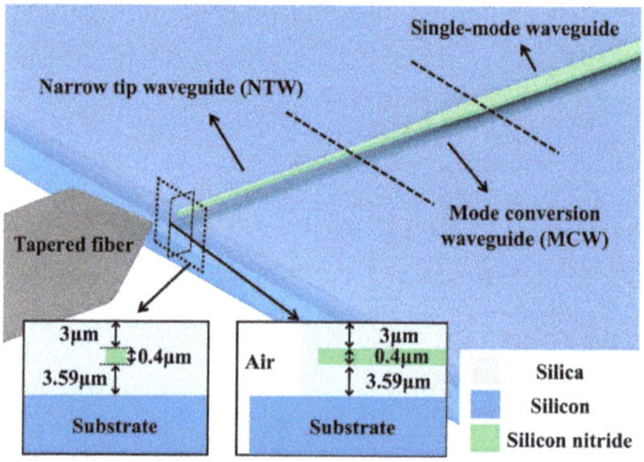

Figure 2. Schematic of the SiN SSC with three parts.

2.1. *The Design of Narrow Tip Waveguide (NTW)*

To ensure that the SiN SSC has a high coupling efficiency with the III-V and the TFLN waveguides (see Figure 1), it is necessary that the MFD of its tip (i.e., the NTW in Figure 2) matches that of these heterogeneous waveguides. To facilitate the measurement, we used tapered fibers or high-numerical aperture (NA) fibers with MFDs of 2.5 μm, 4.5 μm, and 6.5 μm to mimic the heterogeneous waveguides. In the following design, we chose the target MFD of 4.5 μm.

It should be noted that the tapered fiber in the following measurements can easily access the SSC with a protruding silicon substrate (see the second inset in Figure 2), which is the result of ~100 μm deep Si etching. This is not the case for III–V and TFLN chips with flat facets. Therefore, we designed a long NTW region of the same width for the polishing of the SiN-on-SOI chip, which could maintain the tip width after polishing, thus maintaining the chip MFD.

A numerical analysis was conducted to determine the mode overlap between the fiber with a 4.5 μm MFD and NTW modes for various waveguide widths, as visualized in Figure 3. The simulation tools used in this work were Lumerical FDTD or MODE solvers. The results indicate that optimal mode overlap is achieved at a waveguide width of 240 nm. However, the technological limitations of our current fabrication process restricted the minimum achievable waveguide width to 250 nm. Thus, a waveguide width of 250 nm was selected for the NTW.

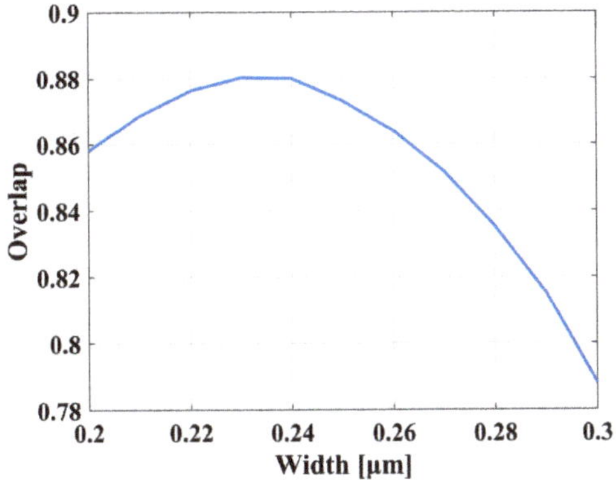

Figure 3. Mode overlapping integral results vary with the NTW width.

2.2. The Design of the Mode Conversion Waveguide (MCW)

Given the narrow waveguide width at the SSC tip and the wider waveguide width required for the low-loss propagation of silicon nitride, a tapered waveguide design was employed to facilitate a smooth mode transition. Numerical simulations were performed to assess the mode conversion efficiency as a function of taper length, as depicted in Figure 4. The simulation results reveal that a mode conversion efficiency of 99% can be attained with a taper length of approximately 771 μm. Even relaxing the efficiency requirement to 97% still necessitates a taper length of 359 μm. Such significant taper lengths would result in a substantial device footprint, hindering the realization of high-density integration.

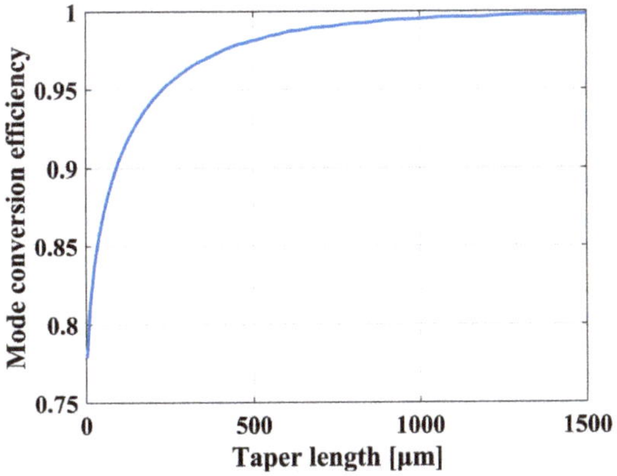

Figure 4. Mode conversion efficiency varies with taper length (linear taper with the width varying from 250 nm to 1000 nm).

To pinpoint the source of taper loss and explore potential optimizations, we conducted a numerical simulation. A linear taper with an initial width of 625 nm (the midpoint between 250 nm and 1000 nm) and a final width of 1000 nm was analyzed for varying lengths. The results, illustrated in Figure 5, indicate that a taper length of no more than 20 μm is sufficient to achieve adiabatic mode transport. Remarkably, even with a taper length of only 1 μm, the mode conversion efficiency exceeded 99%. This suggests that the dominant loss mechanism in a linear taper transitioning from 250 nm to 1000 nm originates from the narrower regions. Consequently, a segmented taper design, allocating greater lengths to the narrower sections, could potentially achieve adiabatic conversion with a significantly shorter overall length compared to the 771 μm taper (~99% efficiency) given in Figure 4.

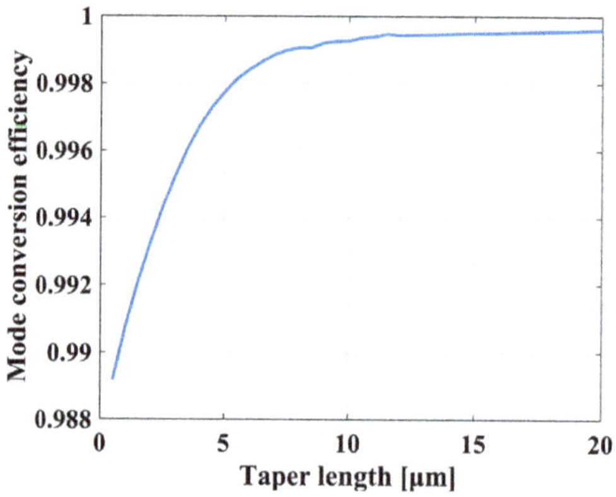

Figure 5. Mode conversion efficiency varies with taper length (linear taper with the width varying from 625 nm to 1000 nm).

To optimize the taper design, a systematic approach was employed to determine the optimal number of segments and the width of each segment. The design methodology aimed to balance the need for rapid transition to the wider waveguide sections with the requirement for gradual transitions in the narrower regions to minimize losses. Consequently, an initial width of 250 nm was followed by a first segment with an end width of either 300 nm or 350 nm. Subsequently, the second segment was designed with an end width of 400 nm, 500 nm, or 600 nm. At this point, the waveguide width was sufficiently large to allow for a direct transition to the target width of 1000 nm. A comprehensive simulation analysis was conducted for each configuration, as depicted in Figure 6.

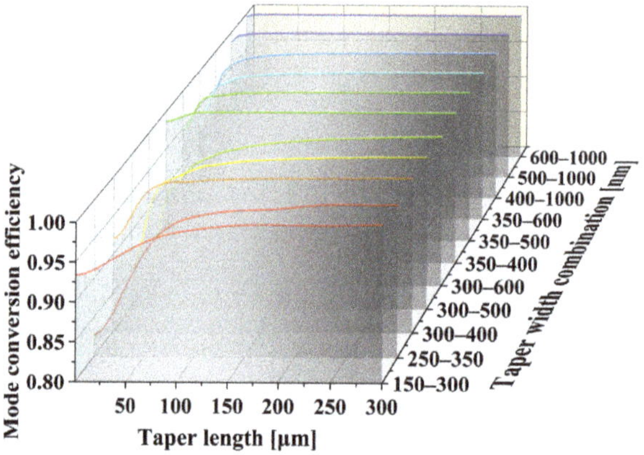

Figure 6. Mode conversion efficiency of different taper width combinations varies with taper length.

To further optimize the taper design, a comprehensive analysis of various combinations was undertaken. The primary objective was to identify a configuration that not only minimized the device footprint but also maintained low-loss performance. For tapers with narrow initial waveguide widths, a slight relaxation of the mode conversion efficiency requirement to 97% was acceptable to achieve a more compact design. Conversely, for regions with wider initial waveguide widths, a more stringent mode conversion efficiency > 99%

was imposed to minimize the overall loss. The detailed results of this optimization process are presented in Figure 7.

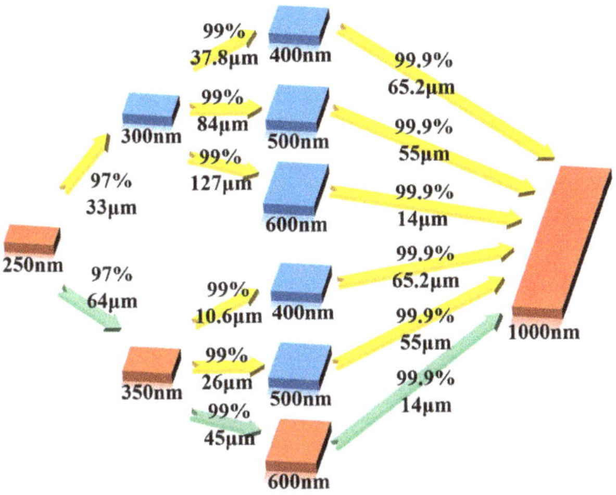

Figure 7. Length and mode conversion efficiency of taper combinations with different start/end widths. Final choice is indicated by orange rectangles and green arrows.

A comprehensive analysis of the simulation results led to the identification of the optimal taper width combination of 250-350-600-1000 nm. This configuration exhibits the smallest device footprint while maintaining a high efficiency. The selected results are highlighted by orange rectangles and green arrows in Figure 7. A simulation of the optimized structure was performed, as illustrated in Figure 8. The electric field distributions of the TM and TE modes for different locations are presented in Figure 8(a1–a4) and Figure 8(b1–b4), respectively. The corresponding positions of these modes are indicated in Figures 8a and 8b. As observed in Figure 8c, the coupling losses for the TE and TM modes are less than 0.56 dB and 1.01 dB, respectively.

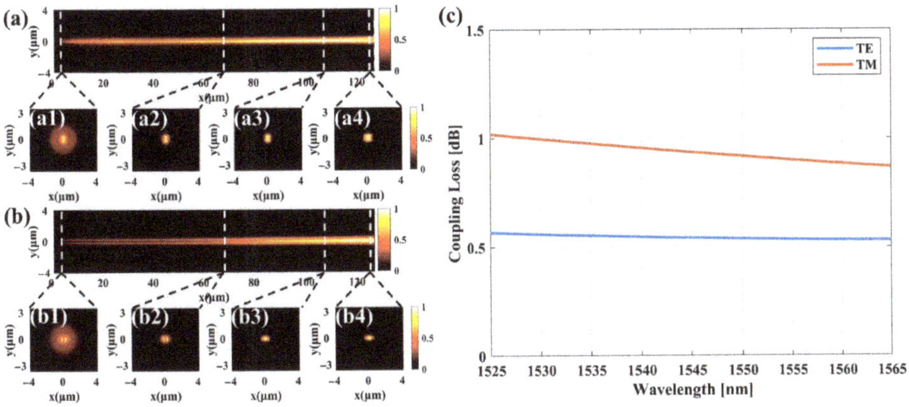

Figure 8. Simulated electric field distributions of (**a**) TM and (**b**) TE mode. (**a1–a4**) Fundamental TM mode at different cross-sections along the SSC. (**b1–b4**) Fundamental TE mode at different cross-sections along the SSC. (**c**) TE and TM mode coupling loss in C-band of designed SSC.

To assess the robustness of the designed device against fabrication variations, a sensitivity analysis was conducted. The coupling efficiency of the device was evaluated for waveguide width variations of up to ±125 nm, as depicted in Figure 9. The results indicate that the TE mode coupling efficiency remained within 1 dB for width variations ranging from −87 nm to +103 nm. Similarly, the TM mode coupling efficiency was maintained within 1 dB for width variations between −125 nm and +96 nm. These findings demonstrate the tolerance of the device to fabrication imperfections.

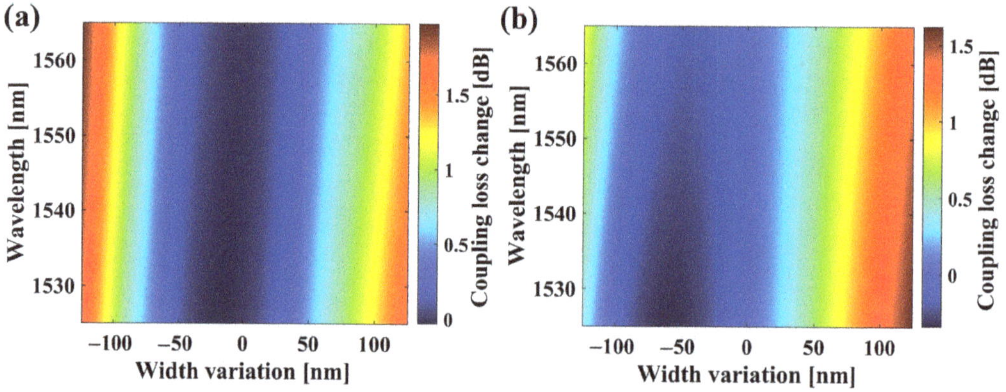

Figure 9. (**a**) The impact of width variation on the coupling efficiency of the coupler for the (**a**) TE mode and (**b**) TM mode.

3. Experimental Results

The designed SSC was fabricated using 193 nm photo-lithography on a 200 mm SOI wafer in a commercial silicon photonics foundry (Institute of Microelectronics, Chinese Academy of Sciences). The microscope image is presented in Figure 10. More specific experimental details are provided in Appendix B.

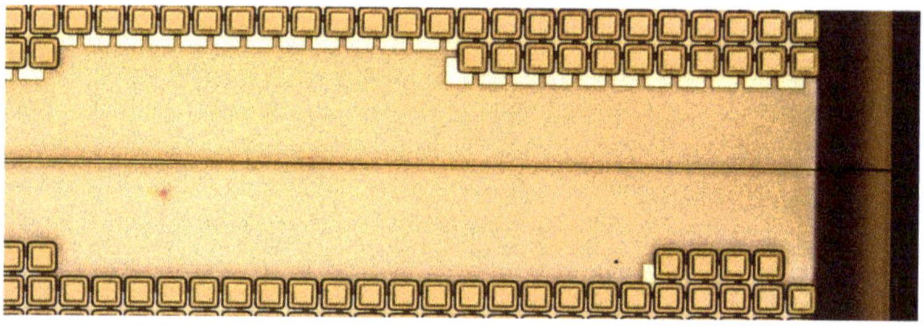

Figure 10. A microscope photo of the designed SSC.

A PDL test was conducted on the structure. Tapered fibers with MFD of 4.5 μm were used for our tests. The measured and simulated results are presented in Figure 11. The 100 μm NTW resulted in an increase in the overall device loss compared to the previous simulation result. Therefore, a re-simulation was performed with the 100 μm NTW. As can be seen in the figure, the measured (simulated) coupling losses of the TE and TM modes in the C-band are less than 0.9 (0.81) dB and 1.38 (1.36) dB, respectively.

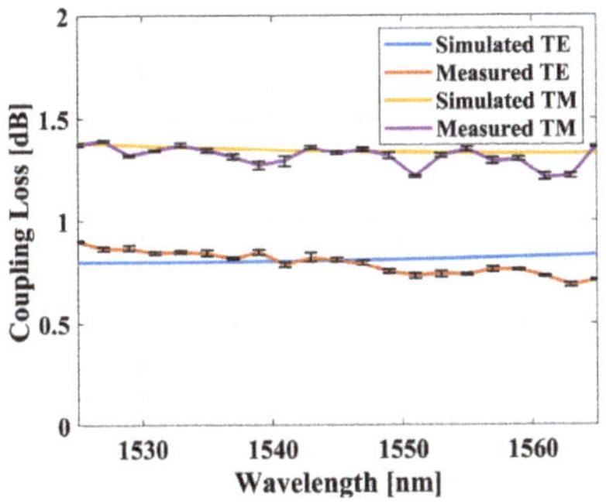

Figure 11. Simulated and measured results of TE/TM coupling loss.

To evaluate the device's sensitivity to alignment errors, an alignment tolerance analysis was performed. The results, as shown in Figure 12, indicate that the vertical and horizontal 3 dB alignment tolerances were 3.6 μm and 3.4 μm, respectively.

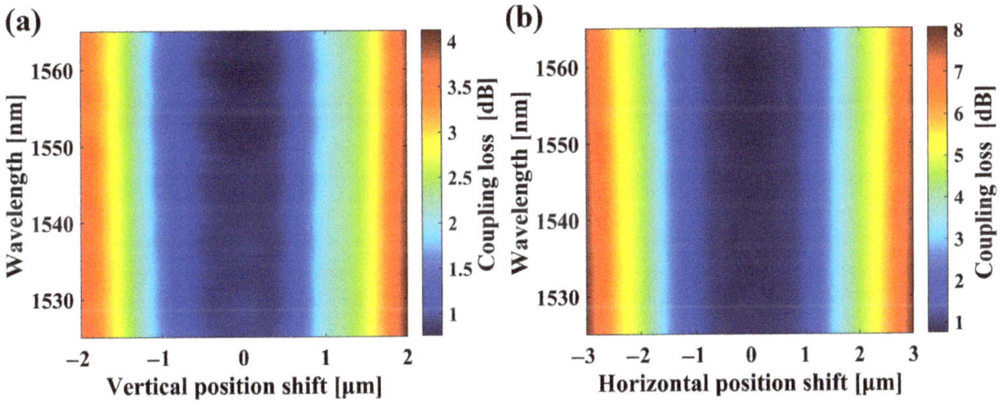

Figure 12. (**a**) The vertical and (**b**) horizontal alignment tolerances of the designed SSC.

The SSC was characterized over an input optical power range of 10 mW to 1 W. As depicted in Figure 13, the device exhibited a relatively constant loss profile across this power range. At 1W input power, the additional loss was approximately 0.1 dB.

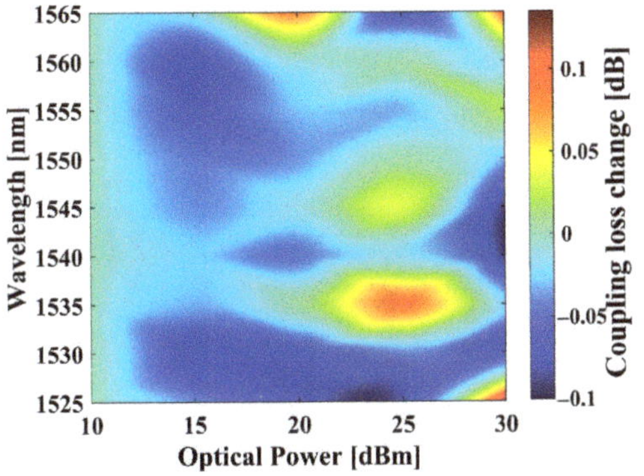

Figure 13. Variation of coupling loss of designed SSC over 1 W input power range.

To further assess the impact of fiber MFD, additional measurements were performed using fibers with MFDs of 2.5 μm and 6.5 μm. The 6.5 μm high-NA fiber was packaged in a V-groove using a refractive-index-matching fluid prior to testing. The corresponding measurement results are presented in Figure 14.

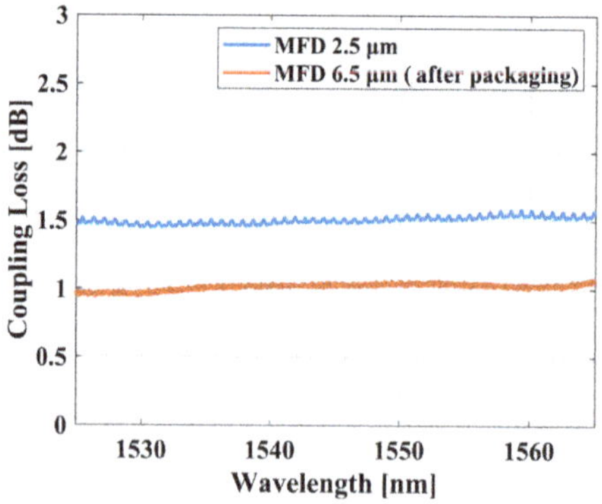

Figure 14. Measurement results for fibers with mode field diameters of 2.5 μm and 6.5 μm.

4. Discussion

The proposed structure can be prepared in a relatively straightforward and versatile process, which can be implemented using standard CMOS processes with a minimum process size of 250 nm. In the absence of any additional manufacturing processes, such as the incorporation of air slots, the cost associated with the packaging of the SSC will be minimized. Here, we make a comprehensive comparison between previously reported SSCs and the proposed structure, as shown in Table 1. It can be clearly seen that only our structure is capable of adapting to a wide range of fiber MFDs. Meanwhile, our structure has a small footprint, low loss, and is insensitive to polarization compared to other works. In addition, it is worth mentioning that our structure is the only one that has been tested for stability under 1W optical input power and successfully verified its excellent stability performance.

Table 1. A comparison with the state of the art.

Refs.	Material	Optical Power	Length (μm)	Structure	MFD (μm)	Operating Wavelength (nm)	Loss (dB)	
							TE	TM
[19]	SiN	-	76	Muti-stage taper	4	1530–1630	0.36	0.55
[32]	SiN	-	500	Inverted taper	4.1	1450–1650	0.45	-
[33]	SiN	-	150	Inverted lateral taper	2.5	1525–1565	1.47	-
[34]	SiON-SiN	- -	780	Inverted taper	10.4	1260–1360 1500–1600	0.64 1.49	0.7 1.55
[35]	SiN	-	-	Air trench and inverted taper	10.4	1500–1630	0.85	1.09
This work	SiN	1W	123	Muti-stage taper	2.5–6.5	1525–1565	0.9	1.38

5. Conclusions

This paper presents a comprehensive design methodology for a compact silicon nitride SSC on an SOI wafer using standard silicon photonics processes. We add a straight narrow waveguide at the front end of the SSC to facilitate dicing and polishing. The designed structure has a small footprint and exhibits low losses, with both dual-polarization losses within 1.5 dB in the C-band for a tapered fiber with a 4.5 μm MFD. We evaluated the influence of fabrication imperfections, alignment errors, and high-power optical input. The proposed structure exhibits robust performance. In particular, the SSC exhibits < 1.5 dB coupling loss with fibers having MFDs of 2.5 μm and 6.5 μm. This makes it a promising candidate for the construction of III–V, TFLN, and SiN-on-SOI electro-optic chiplets capable of handling high-power optical input.

Author Contributions: Conceptualization, E.Z. and L.Z.; methodology, E.Z.; software, E.Z.; validation, E.Z. and Y.Z.; formal analysis, E.Z. and L.Z.; investigation, E.Z. and L.Z.; resources, L.Z. and X.Y.; data curation, E.Z. and Y.Z.; writing—original draft preparation, E.Z.; writing—review and editing, E.Z. and L.Z.; visualization, E.Z.; supervision, L.Z. and X.Y.; project administration, L.Z. and X.Y.; funding acquisition, L.Z. and X.Y. All authors have read and agreed to the published version of the manuscript.

Funding: This research was funded by the fund of State Key Laboratory of Information Photonics and Optical Communications (Beijing University of Posts and Telecommunications) (No. IPOC2024ZZ03), P. R. China and BUPT Excellent Ph.D. Students Foundation (CX20242039).

Institutional Review Board Statement: Not applicable.

Informed Consent Statement: Not applicable.

Data Availability Statement: All relevant data to interpret the results of this study are included in the figures. All other data that support the plots within this paper and other findings of this study are available from the corresponding author upon reasonable request.

Conflicts of Interest: The authors declare no conflicts of interest.

Appendix A

Table A1. All abbreviations used in the manuscript.

Full Name	Abbreviation
microwave photonics	MWP
signal-to-noise ratio	SNR
spot-size converter	SSC
two-photon absorption	TPA
silicon nitride	SiN
silicon-on-insulator	SOI
mode field diameter	MFD
photonics integrated circuit	PIC
complementary metal oxide semiconductor	CMOS
thin-film lithium niobate	TFLN
grating coupler	GC
polarization-dependent loss	PDL
silicon oxynitride	SiON
narrow tip waveguide	NTW
mode conversion waveguide	MCW
numerical aperture	NA

Appendix B

In our experiments, we employed a Golight laser, erbium-doped fiber amplifier, and detector. During high-power testing, we utilized a fixed 30 dB loss attenuator to safeguard the detector from excessive optical power. Furthermore, the precise alignment of the optical fiber and chip was achieved using a high-precision six-axis stage, as depicted in Figure A1.

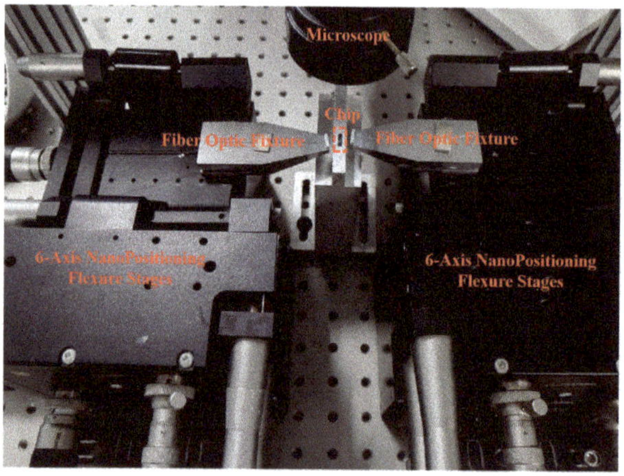

Figure A1. Experimental setup.

The specific testing procedure is outlined as follows.

(1) Connect the laser, EDFA, polarization controller, and photodetector, and secure the optical fibers onto the fiber optic fixtures.
(2) Calibrate the EDFA to determine the required current to achieve the target power at different wavelengths. Given the need for high-power testing later, the attenuator should also be calibrated, with the attenuation fixed at 30 dB.
(3) Estimate the placement position of the chip and adjust the positions of the two six-axis stages to ensure sufficient movement range. Once the positions of the six-axis stages are confirmed, secure them with screws. Adjust the microscope position to ensure a clear view of the chip.
(4) Place the chip on the chip table and ensure that the edge of the chip is parallel to the table.
(5) Under low optical power (10 dBm), observe the fiber position through the microscope and the power received by the photodetector simultaneously. Continuously adjust the positions (in the x, y, z directions) and angles of the two fibers until the power received by the photodetector is maximized.
(6) Rotate the polarization controller to maximize the power received by the photodetector, at which point the corresponding loss represents the coupling loss for the TE mode. Rotate the polarization controller again to minimize the power, and the corresponding loss represents the coupling loss for the TM mode.
(7) During high-power testing, it is necessary to first insert an attenuator prior to the photodetector. By adjusting the wavelength of the laser and the output power of the EDFA, and then recording the output of the optical detector, the coupling loss of the SSC under high-optical-power conditions can be accurately determined.

References

1. Shi, Y.; Zhang, Y.; Wan, Y.; Yu, Y.; Zhang, Y.; Hu, X.; Xiao, X.; Xu, H.; Zhang, L.; Pan, B. Silicon photonics for high-capacity data communications. *Photonics Res.* **2022**, *10*, A106–A134. [CrossRef]
2. Doerr, C.R. Silicon photonic integration in telecommunications. *Front. Phys.* **2015**, *3*, 37. [CrossRef]
3. Silverstone, J.W.; Bonneau, D.; O'Brien, J.L.; Thompson, M.G. Silicon Quantum Photonics. *IEEE J. Sel. Top. Quantum Electron.* **2016**, *22*, 390–402. [CrossRef]
4. Rumley, S.; Nikolova, D.; Hendry, R.; Li, Q.; Calhoun, D.; Bergman, K. Silicon photonics for exascale systems. *J. Light. Technol.* **2015**, *33*, 547–562. [CrossRef]

5. Zilkie, A.J.; Srinivasan, P.; Trita, A.; Schrans, T.; Yu, G.; Byrd, J.; Nelson, D.A.; Muth, K.; Lerose, D.; Alalusi, M.; et al. Multi-Micron Silicon Photonics Platform for Highly Manufacturable and Versatile Photonic Integrated Circuits. *IEEE J. Sel. Top. Quantum Electron.* **2019**, *25*, 8200713. [CrossRef]
6. Aalto, T.; Cherchi, M.; Harjanne, M.; Bhat, S.; Heimala, P.; Sun, F.; Kapulainen, M.; Hassinen, T.; Vehmas, T. Open-Access 3-μm SOI Waveguide Platform for Dense Photonic Integrated Circuits. *IEEE J. Sel. Top. Quantum Electron.* **2019**, *25*, 1–9. [CrossRef]
7. Guo, Y.; Li, X.; Jin, M.; Lu, L.; Xie, J.; Chen, J.; Zhou, L. Hybrid integrated external cavity laser with a 172-nm tuning range. *APL Photonics* **2022**, *7*, 066101. [CrossRef]
8. Xu, M.; He, M.; Zhang, H.; Jian, J.; Pan, Y.; Liu, X.; Chen, L.; Meng, X.; Chen, H.; Li, Z.; et al. High-performance coherent optical modulators based on thin-film lithium niobate platform. *Nat. Commun.* **2020**, *11*, 3911. [CrossRef] [PubMed]
9. Li, T.; Hou, J.; Yan, J.; Liu, R.; Yang, H.; Sun, Z. Chiplet Heterogeneous Integration Technology—Status and Challenges. *Electronics* **2020**, *9*, 670. [CrossRef]
10. Cheng, L.; Mao, S.; Li, Z.; Han, Y.; Fu, H.Y. Grating Couplers on Silicon Photonics: Design Principles, Emerging Trends and Practical Issues. *Micromachines* **2020**, *11*, 666. [CrossRef] [PubMed]
11. Liu, L.; Pu, M.; Yvind, K.; Hvam, J.M. High-efficiency, large-bandwidth silicon-on-insulator grating coupler based on a fully-etched photonic crystal structure. *Appl. Phys. Lett.* **2010**, *96*, 051126. [CrossRef]
12. Halir, R.; Cheben, P.; Janz, S.; Xu, D.X.; Molina-Fernández, Í.; Wangüemert-Pérez, J.G. Waveguide grating coupler with subwavelength microstructures. *Opt. Lett.* **2009**, *34*, 1408–1410. [CrossRef] [PubMed]
13. Taillaert, D.; Bienstman, P.; Baets, R. Compact efficient broadband grating coupler for silicon-on-insulator waveguides. *Opt. Lett.* **2004**, *29*, 2749–2751. [CrossRef]
14. Wang, Y.; Flueckiger, J.; Lin, C.; Chrostowski, L. Universal grating coupler design. *Proc. Photonics N. 2013 Int. Soc. Opt. Photonics* **2013**, *8915*, 8201109.
15. Taillaert, D.; Van Laere, F.; Ayre, M.; Bogaerts, W.; Van Thourhout, D.; Bienstman, P.; Baets, R. Grating Couplers for Coupling between Optical Fibers and Nanophotonic Waveguides. *Jpn. J. Appl. Phys.* **2006**, *45*, 6071–6077. [CrossRef]
16. Marchetti, R.; Lacava, C.; Khokhar, A.; Chen, X.; Cristiani, I.; Richardson, D.J.; Reed, G.T.; Petropoulos, P.; Minzioni, P. High-efficiency grating-couplers: Demonstration of a new design strategy. *Nat. Photonics* **2013**, *7*, 16670. [CrossRef] [PubMed]
17. Tang, Y.; Wang, Z.; Wosinski, L.; Westergren, U.; He, S. Highly efficient nonuniform grating coupler for silicon-on-insulator nanophotonic circuits. *Opt. Lett.* **2010**, *35*, 1290–1292. [CrossRef] [PubMed]
18. Mekis, A.; Gloeckner, S.; Masini, G.; Narasimha, A.; Pinguet, T.; Sahni, S.; De Dobbelaere, P. A Grating-Coupler-Enabled CMOS Photonics Platform. *IEEE J. Sel. Top. Quantum Electron.* **2011**, *17*, 597–608. [CrossRef]
19. Bhandari, B.; Im, C.S.; Lee, K.P.; Kim, S.M.; Oh, M.C.; Lee, S.S. Compact and Broadband Edge Coupler Based on Multi-Stage Silicon Nitride Tapers. *IEEE Photonics J.* **2020**, *12*, 6602511. [CrossRef]
20. Liu, W.; Zhang, J.; Liu, L.; Dai, D.; Shi, Y. High Efficiency Silicon SSC Based On Uniform Arrayed Waveguides With Un-Patterned Cladding. *IEEE Photonics Technol. Lett.* **2020**, *32*, 1077–1080. [CrossRef]
21. Mu, X.; Wu, S.; Cheng, L.; Fu, H. SSCs in Silicon Photonic Integrated Circuits: A Review. *Appl. Sci.* **2010**, *10*, 1538. [CrossRef]
22. Wohlfeil, B.; Rademacher, G.; Stamatiadis, C.; Voigt, K.; Zimmermann, L.; Petermann, K. A two-dimensional fiber grating coupler on SOI for mode division multiplexing. *IEEE Photonics Technol. Lett.* **2016**, *28*, 1241–1244. [CrossRef]
23. Lai, Y.; Yu, Y.; Fu, S.; Xu, J.; Shum, P.P.; Zhang, X. Compact double-part grating coupler for higher-order mode coupling. *Opt. Lett.* **2018**, *43*, 3172–3175. [CrossRef] [PubMed]
24. Papes, M.; Cheben, P.; Benedikovic, D.; Schmid, J.H.; Pond, J.; Halir, R.; Ortega-Moñux, A.; Wangüemert-Pérez, G.; Ye, W.N.; Xu, D.X. Fiber-chip SSC with large mode size for silicon photonic wire waveguides. *Opt. Express* **2016**, *24*, 5026–5038. [CrossRef] [PubMed]
25. Wang, X.; Quan, X.; Liu, M. Cheng Silicon-Nitride-Assisted SSC Interfacing With High Numerical Aperture Fiber. *IEEE Photonics Technol. Lett.* **2019**, *31*, 349–352. [CrossRef]
26. He, A.; Guo, X.; Wang, K.; Zhang, Y.; Su, Y. Low Loss, Large Bandwidth Fiber-Chip SSCs Based on Silicon-on-Insulator Platform. *J. Lightwave Technol.* **2020**, *38*, 4780–4786. [CrossRef]
27. Fan, Y.; van Rees, A.; Van der Slot, P.J.; Mak, J.; Oldenbeuving, R.M.; Hoekman, M.; Geskus, D.; Roeloffzen, C.G.; Boller, K.J. Hybrid integrated InP-Si_3N_4 diode laser with a 40-Hz intrinsic linewidth. *Opt. Express* **2020**, *28*, 21713–21728. [CrossRef] [PubMed]
28. Zhu, W.; Deng, C.; Wang, D.; Wang, Q.; Sun, Y.; Wang, J.; Yun, B.; Hu, G.; Cui, Y. Broadband and easily fabricated double-tip edge coupler based on thin-film lithium niobate platform. *Opt. Commun.* **2024**, *573*, 131031. [CrossRef]
29. Tang, Y.; Li, H.; Chen, Q.; Dai, X.; Lu, Q.; Lu, M.; Guo, W. Low Loss Thin-Film Lithium Niobate Modulator with T-Rail Electrodes on Quartz Substrate Fabricated by Photolithography. In Proceedings of the 2023 Conference on Lasers and Electro-Optics (CLEO), San Jose, CA, USA, 7–12 May 2023. Technical Digest Series, Paper STh4O.4.
30. Li, T.; Shang, C.; Wang, X.; Lü, W.; Zhang, Z.; Zeng, C.; Liu, Y.; Xia, J. Low-loss and broadband complementary dual-output electro-optic modulator based on thin-film lithium niobate. *Chin. Opt. Lett.* **2024**, *22*, 092201–922016. [CrossRef]

31. Cheng, Y.; Li, X.; Feng, L.; Li, H.; Sun, W.; Song, X.; Ding, Y.; Guo, G.; Wang, C. Cryogenic nonlinear conversion processes in periodically-poled thin-film lithium niobate waveguides. *arXiv* **2024**, arXiv:2408.05907. [CrossRef]
32. Zhu, T.; Hu, Y.; Gatkine, P.; Veilleux, S.; Bland-Hawthorn, J.; Dagenais, M. Ultrabroadband high coupling efficiency fiber-to-waveguide coupler using Si_3N_4 /SiO_2 waveguides on silicon. *IEEE Photonics J.* **2016**, *8*, 1–12.
33. Fernández, J.; Baños, R.; Doménech, D.; Domínguez, C.; Muñoz, P. Low-loss inverted taper edge coupler in silicon nitride. *IET Optoelectron.* **2019**, *13*, 62–66. [CrossRef]
34. Li, Y.; Xu, J.; Quan, X.; Guo, C.; Jin, X.; Cheng, X. O-band and C-band dual-polarization SMF-28 edge coupler with SiON taper cladding based on silicon nitride platform. *Opt. Express* **2024**, *32*, 28259–28266. [CrossRef]
35. Liang, Y.; Li, Z.; Fan, S.; Feng, J.; Liu, D.; Liao, H.; Yang, Z.; Feng, J.; Cui, N. Ultra-low loss SiN edge coupler interfacing with a single-mode fiber. *Opt. Lett.* **2022**, *47*, 4786–4789. [CrossRef] [PubMed]
36. Kohli, N.; Ménard, M.; Ye, W.N. Efficient TE/TM spot-size converter for broadband coupling to single mode fibers. *OSA Contin.* **2019**, *2*, 2428–2438. [CrossRef]
37. Huang, H.; Shi, Z.; Talli, G.; Kuschnerov, M.; Penty, R.; Cheng, Q. Photonic chiplet interconnection via 3D-nanoprinted interposer. *Light. Adv. Manuf.* **2024**, *5*, 46. [CrossRef]

photonics

MDPI

Review

Photonics-Enabled High-Sensitivity and Wide-Bandwidth Microwave Phase Noise Analyzers

Jingzhan Shi, Baojin Tu and Yiping Wang *

Ministry of Education Key Laboratory of NSLSCS, School of Computer and Electronic Information, Nanjing Normal University, Nanjing 210016, China; shijz@nnu.edu.cn (J.S.); 232212043@njnu.edu.cn (B.T.)
* Correspondence: ypwang@njnu.edu.cn

Abstract

Phase noise constitutes a pivotal performance parameter in microwave systems, and the evolution of microwave signal sources presents new demands on phase noise analyzers (PNAs) regarding sensitivity and bandwidth. Traditional electronics-based PNAs encounter significant limitations in meeting these advanced requirements. This paper provides an overview of recent progress in photonics-based microwave PNA research. Microwave photonic (MWP) PNAs are categorized into two main types: phase-detection-based and frequency-discrimination-based architectures. MWP phase-detection-based PNAs utilize ultra-short-pulse lasers or optical–electrical oscillators as reference sources to achieve superior sensitivity. On the other hand, MWP frequency-discrimination-based PNAs are further subdivided into photonic-substitution-type PNA and MWP quadrature-frequency-discrimination-based PNA. These systems leverage innovative MWP technologies to enhance overall performance, offering broader bandwidth and higher sensitivity compared to conventional approaches. Finally, the paper addresses the current challenges faced in phase noise measurement technologies and suggests potential future research directions aimed at improving measurement capabilities.

Keywords: phase noise analyzer; microwave photonics; high sensitivity; wide bandwidth; phase detection; frequency discrimination

1. Introduction

Phase noise serves as a quantitative measure of short-term frequency stability in microwave signals [1,2]. Specifically, it is mathematically defined as half of the single-sided power spectral density (PSD) associated with random phase fluctuations in a microwave signal, expressed as [3]:

$$\mathcal{L}(f_{\mathrm{m}}) \equiv \frac{S_{\varphi}(f_{\mathrm{m}})}{2} \tag{1}$$

where f_{m} represents the frequency offset from the carrier, and $S_{\varphi}(f_{\mathrm{m}})$ denotes the single-sided PSD of the phase fluctuations $\varphi(t)$.

Phase noise constitutes a pivotal performance parameter in microwave engineering systems such as radar systems [4], timing devices [5], and communication equipment [6]. The precise measurement of this parameter not only provides theoretical foundations for system development but also serves as a critical technical basis for fault diagnosis and performance enhancement, thereby emerging as a fundamental challenge in contemporary microwave engineering research.

https://doi.org/10.3390/photonics12070691

To meet the increasingly stringent requirements of various application systems for microwave signal sources, researchers worldwide have dedicated significant efforts to reducing phase noise and expanding frequency ranges. As the phase noise performance of microwave signal sources continues to improve, the generated phase noise intensity becomes increasingly weaker, necessitating phase noise analyzers (PNAs) with enhanced sensitivity to accurately detect these extremely weak noise signals. During the development and testing of low-noise signal sources, insufficient measurement system sensitivity may lead to an inability to distinguish between the inherent low phase noise of the signal source and the noise introduced by the measurement system itself, resulting in measurement deviations. For instance, the X-band optoelectronic oscillator (OEO) produced by the OEwaves Company exhibits remarkable phase noise performance of −163 dBc/Hz @ 6 kHz [7], presenting significant challenges to the measurement sensitivity of PNA. Furthermore, as microwave signal sources evolve towards broadband and high-frequency applications, PNAs must possess sufficient measurement bandwidth to cover all operational frequency bands. With the emergence of new technologies such as 5G communications, millimeter-wave radar, and satellite communications, the operational frequencies of microwave signal sources continue to extend into higher bands. In the millimeter-wave frequency range of 5G communications (e.g., 24.25 GHz to 52.6 GHz), signal sources must provide stable high-frequency signals to support high-speed data transmission. Meanwhile, millimeter-wave radars tend to adopt 77 GHz, 79 GHz, or even higher frequencies to achieve enhanced resolution and precision. This implies that signal sources must maintain excellent phase noise characteristics at high frequencies, as high-frequency signals exhibit greater sensitivity to phase noise variations, where even minor phase fluctuations can significantly impact signal quality. For microwave signal sources with rapid switching capabilities, corresponding phase noise measurement systems must execute measurements swiftly to enable real-time monitoring of phase noise variations during frequency transitions. Particularly in testing frequency-hopping communication signal sources, where frequent frequency changes occur, inadequate measurement speed may result in missed detection of phase noise conditions during each frequency transition, thereby hindering timely problem identification. In summary, the developmental trends of microwave signal sources towards improved stability, broader bandwidths, and faster switching speeds present new challenges for PNAs regarding sensitivity, bandwidth, and speed. These emerging demands will conversely drive advancements in microwave signal source technology, enabling better adaptation to the growing high-performance requirements of modern communication, radar, and electronic measurement industries.

However, due to the inherent limitations of electronic components in terms of speed and sensitivity, conventional electronics-based PNAs are increasingly challenged in meeting the growing demands for ultra-wideband and high-sensitivity measurement capabilities. This electronic bottleneck not only constrains the performance enhancement of such instruments but also impedes their broad applications in high-frequency bands and complex signal environments. Consequently, the development of novel phase noise measurement technologies and instrumentation has emerged as one of the key research priorities in contemporary metrology. Fortunately, the rapid development of microwave photonics (MWP) in recent years has brought novel approaches to high-performance phase noise measurement [8,9]. By leveraging the inherent advantages of photonic technologies such as broadband capability, low loss, low jitter, and electromagnetic interference immunity, high-sensitivity and wide-bandwidth microwave phase noise measurement can be effectively achieved [10,11].

This paper reviews recent advancements in photonics-based microwave PNA research, with an overall summary diagram presented in Figure 1. The paper is organized as follows. Section 2 elucidates the fundamental principles, advantages, and limitations of typical phase noise measurement methodologies. Section 3 provides a comprehensive analysis of MWP phase-detection-based PNA implementations. Section 4 examines recent developments in MWP frequency-discrimination-based PNA systems. Finally, Section 5 discusses current challenges in phase noise measurement technologies and proposes potential research directions for performance enhancement.

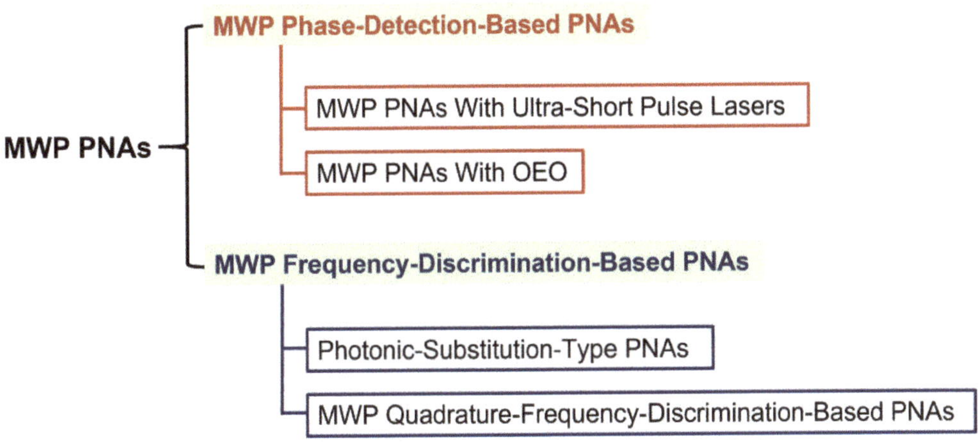

Figure 1. Overall summary diagram of photonics-based microwave PNAs.

2. Basic Phase Noise Measurement Methods

There are mainly four kinds of basic phase noise measurement methods [12,13], including the direct spectrum method, the phase detection method, the frequency discrimination method, and the dual-channel cross-correlation method.

2.1. Direct Spectrum Method

The direct spectrum method represents the most straightforward technique for phase noise measurement, with its fundamental principle depicted in Figure 2. In this approach, the signal under test (SUT) is directly fed into a spectrum analyzer to characterize its spectral properties. The phase noise of the signal source is subsequently determined through the following equation:

$$\mathcal{L}(f_{\mathrm{m}}) = \frac{P_{\mathrm{SSB}}(f_{\mathrm{m}})}{P_{\mathrm{c}}} \tag{2}$$

where $P_{\mathrm{SSB}}(f_{\mathrm{m}}) = S_{\mathrm{v}}(f_{\mathrm{c}} + f_{\mathrm{m}})$ denotes the single-sided PSD of the SUT at an frequency offset f_{m} from the carrier frequency f_{c}, while P_{c} represents the carrier power. The phase noise calculated using Equation (2) approximates the theoretical definition in Equation (1) when two critical conditions are satisfied [3]:

(1) Negligible amplitude noise in the SUT;
(2) Compliance with the small-angle approximation, requiring: root mean square value of the phase fluctuations $\varphi(t)$ is less than 0.01 rad, and peak magnitude of $\varphi(t)$ do not exceed 0.2 rad.

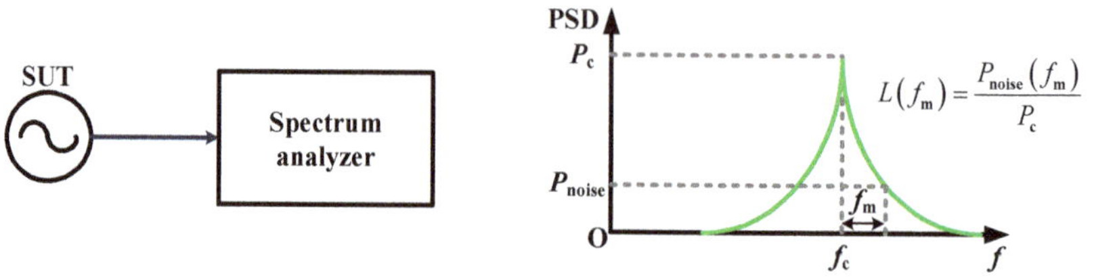

Figure 2. Principle of direct spectrum method. SUT: Signal under test.

This direct spectrum method features a simple structure and easy implementation. However, it has several limitations. Firstly, the method relies on approximate calculations based on Equation (2), making it only suitable for signals with relatively low phase noise and inadequate for microwave signals exhibiting poor phase noise performance. Secondly, the technique cannot differentiate between phase noise and amplitude noise in the measured signal. Consequently, the measurement results encompass both types of noise, leading to inaccuracies in the final phase noise assessment. Thirdly, both the measurement bandwidth and sensitivity of this method are constrained by the operational bandwidth and dynamic range limitations of the spectrum analyzer.

The current mainstream commercial PNAs using direct spectrum method in the market primarily include Keysight's N9030A PXA series signal analyzers, sourced from Keysight Technologies in Santa Clara, CA, USA, and Rohde & Schwarz's FSV series signal analyzers, sourced from Rohde & Schwarz in Munich, Germany. Among these, the N9030A-550 model with the largest measurement bandwidth in the N9030A PXA series features a wideband measurement capability of 3 Hz to 50 GHz. Under 1 GHz carrier frequency conditions, its typical sensitivity reaches −129 dBc/Hz @ 10 kHz. Meanwhile, the flagship model FSV-K40 of the FSV series extends its operational bandwidth to 10 MHz–40 GHz, achieving a sensitivity specification of −110 dBc/Hz @ 10 kHz @ 1 GHz. Notably, although both instruments demonstrate excellent phase noise measurement sensitivity at 1 GHz carrier frequency, their performance exhibits significant degradation as carrier frequency increases. Experimental data reveal that when the test signal carrier frequency is elevated to 38 GHz, the sensitivity of FSV-K40 at 10 kHz offset deteriorates to −92 dBc/Hz, showing an 18 dB performance degradation compared to the measurement at 1 GHz.

2.2. Phase Detection Method

A schematic diagram of the phase detection method principle is shown in Figure 3. This method requires the reference source and the microwave SUT to share identical frequencies while maintaining a 90° phase difference. Both signals are input into a phase detector for phase-to-voltage conversion. After obtaining the PSD of the voltage signal through a spectrum analyzer, calibration is required to eliminate the influence of the phase detector's conversion coefficient, ultimately yielding the PSD of the phase difference between the two signals. Since the phase noise of the reference signal is typically significantly lower than that of the SUT, this PSD directly characterizes the phase fluctuations characteristics of the SUT, which can be converted to phase noise using Equation (1).

Although the phase detection method offers high measurement accuracy, it presents notable implementation constraints: Firstly, obtaining a reference source with identical frequency and superior phase noise performance to the SUT proves challenging in practical measurements, resulting in system measurement bandwidth and sensitivity being limited by the reference source's tunable frequency range and phase noise characteristics, respectively. Secondly, this method is unsuitable for free-running oscillators or sources with poor frequency stability due to difficulties in maintaining the required equal frequency and quadrature phase.

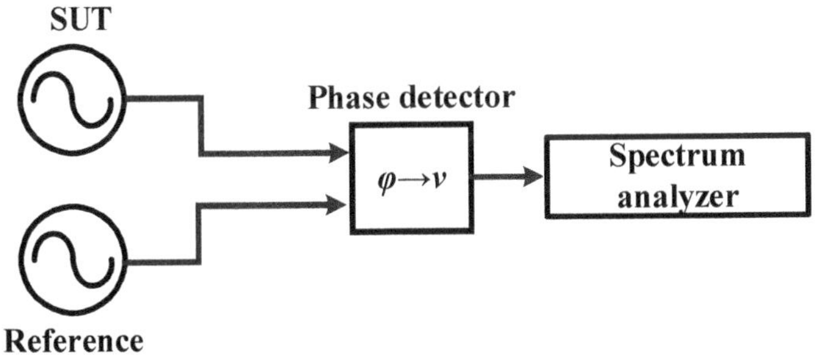

Figure 3. Schematic diagram of phase detection method.

Current mainstream PNAs predominantly adopt phase detection architectures, with representative models including the Keysight E5500 series, sourced from Keysight Technologies in Santa Clara, CA, USA, and NoiseXT PN9000 series (featuring dual phase/frequency detection capabilities), sourced from Voisins-le-Bretonneux, France. Limited by reference source constraints, the E5500 series operates within 50 kHz–26.5 GHz. At lower frequencies (<3 GHz), its typical noise floor reaches −135 dBc/Hz @ 10 kHz, where sensitivity is primarily constrained by reference source phase noise. For high-frequency measurements (18–26.5 GHz), down-conversion via mixers introduces additional noise from local oscillators and mixer components, degrading the noise floor to −118 dBc/Hz @ 10 kHz. In contrast, the PN9000 series extends its measurement range to 2 MHz–50 GHz through multi-stage down-conversion, but similarly exhibits high-frequency sensitivity degradation: achieving −136 dBc/Hz @ 10 kHz at 1 GHz carrier versus −104 dBc/Hz @ 10 kHz at 40 GHz carrier.

2.3. Frequency Discrimination Method

The working principle of the frequency discrimination method is illustrated in Figure 4. The microwave SUT is divided into two paths by a power splitter: one path passes through a phase shifter to the phase detector, while the other is fed to the phase detector via a delay line (time delay τ). The phase shifter is adjusted to maintain a 90° phase difference between the two input signals, enabling the phase detector to output a voltage signal proportional to the phase difference. After obtaining the PSD of this voltage signal through a spectrum analyzer, calibration is required to eliminate the influence of the phase-to-voltage conversion coefficient of the phase detector, ultimately yielding the PSD of the phase difference $\varphi(t) - \varphi(t-\tau)$. The relationship between phase noise $\mathcal{L}(f_m)$ and the single-sideband PSD of phase difference $S_{\Delta\varphi}(f_m)$ can be expressed as:

$$S_{\Delta\varphi}(f_m) = \left|1 - e^{-j2\pi f_m \tau}\right| S_\varphi(f_m) = 8\sin^2(\pi f_m \tau)\mathcal{L}(f_m) \tag{3}$$

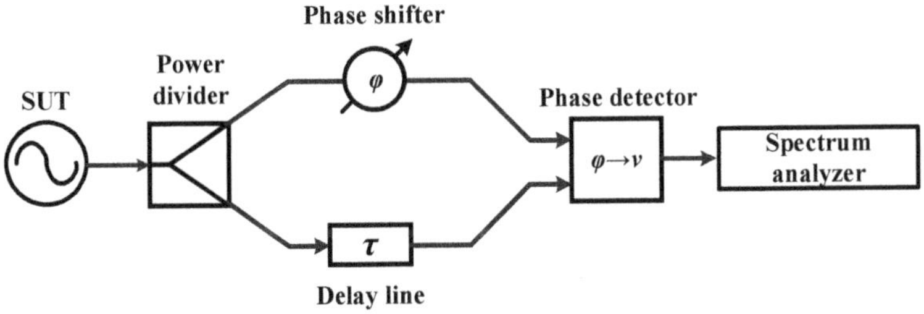

Figure 4. Schematic diagram of frequency discrimination method.

This indicates that phase noise can be obtained by dividing the PSD of the phase difference by $8\sin^2(\pi f_m \tau)$.

Compared with phase detection methods, the frequency discrimination approach eliminates the requirement for a reference source, thereby resolving three inherent limitations of phase detection techniques: restricted measurement bandwidth, compromised sensitivity due to reference source dependency, and lack of applicability to free-running oscillators. However, this method demonstrates diminished measurement accuracy in low-frequency offset regimes. The performance constraints primarily stem from two key delay line parameters:

Sensitivity Trade-off: While delay length exhibits positive correlation with sensitivity, extended electrical delay lines introduce increased attenuation. This attenuation reduces phase detector conversion efficiency, ultimately counteracting the sensitivity enhancement.

Bandwidth Limitations: The operational bandwidth of delay lines fundamentally restricts measurement range, while the finite bandwidth characteristics of electrical phase shifters and phase detectors impose additional performance constraints.

The NoiseXT PN9000 series PNA, a representative dual-mode system integrating both frequency discrimination and phase detection, exemplifies these characteristics. Specifically designed for free-running oscillators or sources with marginal frequency stability, its performance metrics reveal the following: at 1 GHz carrier frequency, the instrument achieves a noise floor of −133 dBc/Hz @ 10 kHz with 100 ns delay, degrading to −124 dBc/Hz @ 10 kHz with 20 ns delay. In high-frequency implementations (e.g., 40 GHz carriers), despite employing low-noise amplifiers and down-conversion modules to compensate for cable attenuation, the noise floor rises to −104 dBc/Hz @ 10 kHz (100 ns delay) and −103 dBc/Hz @ 10 kHz (20 ns delay). This performance degradation underscores the sensitivity compromise introduced by auxiliary circuitry in practical implementations.

2.4. Dual-Channel Cross-Correlation Method

Based on the technical framework of phase detection and frequency discrimination methods, the introduction of dual-channel cross-correlation technology can significantly enhance the sensitivity of PNAs. Figure 5 illustrates the core principle of the dual-channel cross-correlation method: the microwave SUT is equally divided into two paths through a power divider, which are then fed into two structurally identical phase noise measurement channels. Assuming the intrinsic noise of the two channels as $a(t)$ and $b(t)$, respectively, and the phase noise of the measured signal as $c(t)$, the output signals of both channels can be expressed as the superposition of the measured signal's phase noise and their respective system intrinsic noise. By performing cross-power spectral analysis on the two output signals, the mathematical expression is derived as:

$$S_{xy} = \frac{1}{T}E\{(C+A)\times(C+B)^*\} = \frac{1}{T}(E\{CC^*\}+E\{CB^*\}+E\{AC^*\}+E\{AB^*\}) \quad (4)$$

where $A(f)$, $B(f)$, and $C(f)$ represent the Fourier transforms of the corresponding time-domain signals. Based on the statistical independence assumption among noise sources, the expectation values of the last three terms in Equation (4) approach zero, thereby effectively suppressing system intrinsic noise. This technology has been successfully implemented in commercial instruments such as the Rohde & Schwarz FSWP series and NoiseXT's DCNTS/NXA series, significantly improving phase noise measurement sensitivity. Experimental data demonstrate that under 1 GHz carrier frequency and 10 kHz offset conditions, after 1000 cross-correlation processing iterations, the noise floors of FSWP, DCNTS, and NXA systems reach −181 dBc/Hz, −183 dBc/Hz, and −187 dBc/Hz, respectively, showing remarkable improvement compared to traditional single-channel systems (e.g., E5500 and PN9000 series).

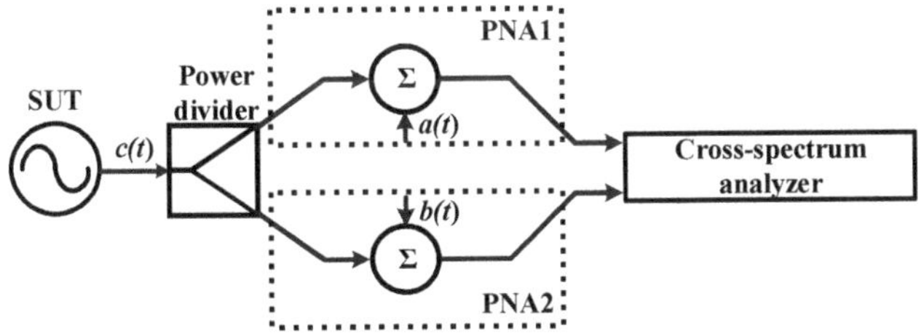

Figure 5. Schematic diagram of dual-channel cross-correlation method.

It should be noted that system sensitivity exhibits positive correlation with cross-correlation processing iterations, while measurement time increases with processing count. Furthermore, this technology introduces significant system architecture complexity while enhancing performance.

3. MWP Phase-Detection-Based PNAs

The analysis of the phase detection method in Section 2.2 indicates that the phase noise of the reference signal source is the primary bottleneck limiting the sensitivity of phase detection. To overcome the limitations imposed by traditional microwave signal sources' phase noise, MWP phase-detection-based PNAs employ ultra-short pulse lasers or OEOs as reference sources, which have superior phase noise characteristics.

3.1. MWP PNA with Ultra-Short-Pulse Lasers

Ultra-short-pulse lasers consist of a sequence of optical pulses with extremely short durations in the time domain, and their unique low time jitter characteristics make them particularly advantageous for high-sensitivity phase noise detection [14–16]. MWP phase detection methods based on this light source measure the phase difference between optical pulses and microwave signals using MWP phase detectors. According to structural features, MWP phase detectors are primarily categorized into Sagnac loop type and dual-output Mach-Zehnder modulator (DO-MZM) type.

Figure 6 illustrates the typical structure of a Sagnac loop MWP phase detector [17–22]. The system uses an ultra-short pulse laser with a repetition frequency f_R as the light source, and the frequency of the microwave SUT is set to Nf_R (where N is an integer), with its phase fluctuations denoted as $\varphi(t)$. This phase detector consists of a circulator, a 50:50 coupler, a Sagnac loop, and a balanced photodetector (as shown in the shaded area of Figure 6). After entering the coupler via the circulator, the optical pulses split into two paths and enter the Sagnac loop. This loop is composed of a phase modulator and a $\pi/2$ phase shifter. Two optical pulses propagate within the Sagnac loop in opposite directions: clockwise and counterclockwise. A $\pi/2$ phase shifter introduces a $\pi/2$ phase shift to one of the optical paths. Additionally, due to the traveling-wave characteristics of the phase modulator, only the forward-propagating optical pulse is modulated by the drive signal, while the backward-propagating pulse remains unmodulated. When applying the SUT as the drive signal to the phase modulator, a phase difference proportional to the microwave signal is introduced between the forward and backward propagating pulses. Specifically, when the zero-crossing point of the microwave signal aligns with the optical pulses, this phase difference becomes approximately proportional to the phase fluctuations $\varphi(t)$, expressed as $k\varphi(t)$ (where k is a constant). The cooperative action of the phase modulator and the phase shifter causes the two paths to generate a composite phase difference of $\Psi(t) = k\varphi(t) + \pi/2$.

After circulating through the loop, the two optical beams interfere at the coupler, and their optical powers satisfy:

$$P_1(t) \propto \cos^2\frac{\Psi(t)}{2},\ P_2(t) \propto \sin^2\frac{\Psi(t)}{2} \quad (5)$$

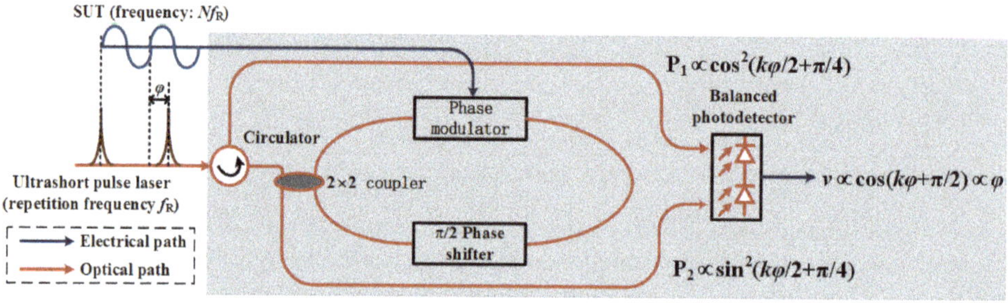

Figure 6. MWP phase detector based on Sagnac loop.

The balanced photodetector measures the power difference $\Delta P(t) \propto \cos\Psi(t)$. Under small phase jitter conditions, $\Delta P(t) \approx k\varphi(t)$, indicating a linear relationship between the output signal and phase fluctuations. By calibrating the system to obtain the phase-to-voltage conversion coefficient, the phase noise of the tested source can be calculated using Equation (1).

As early as 2004, Jungwon Kim et alproposed a MWP phase detector based on an optical fiber Sagnac loop. Experimental results demonstrated that, compared to the phase detector based on a spatial light Sagnac loop, the one based on an optical fiber Sagnac loop achieved a lower noise floor. In 2013, to mitigate the impact of optical pulse amplitude noise on phase detection, the National Physical Laboratory in the UK introduced two adjustable optical attenuators before the balanced photodetector to regulate the power of the two optical pulse signals entering the balanced photodetector, ensuring balanced optical power [20]. However, all these configurations required a $\pi/2$ phase shifter to introduce a $\pi/2$ phase difference between the forward and reverse propagating optical pulses in the Sagnac loop, thus biasing the phase detector at the linear point. The $\pi/2$ phase shifter is temperature-sensitive, leading to reduced stability of the phase detector. To address this issue, Jungwon Kim et al. proposed a balanced MWP phase detector based on the Sagnac loop, which eliminates the need for a $\pi/2$ phase shifter by using a low-frequency signal of $f_R/2$ extracted from ultra-short pulse lasers to regulate the operating point [21,22].

Although the balanced Sagnac-loop-based detector offers improved stability, it requires precise positioning of the phase modulator within the loop. The modulator's position and loop length are dependent on the optical pulse repetition rate, increasing system complexity and tuning difficulty. To overcome these limitations, T. R. Schibli et al. developed a MWP phase detector using DO-MZM [23], as shown in Figure 7. The DO-MZM operates at quadrature bias, where its dual output powers exhibit complementary linear relationships with applied microwave signal amplitude. Through balanced detection, this configuration produces a quasi-linear phase detection characteristic: zero output voltage occurs when microwave signal zero-crossings align perfectly with optical pulses, while phase fluctuations generate proportional voltage outputs. By calibrating the characteristic curve's slope, phase noise can be quantified. To further enhance sensitivity, Schibli's team implemented dual-channel cross-correlation, achieving −167 dBc/Hz @ 10 kHz phase noise measurement sensitivity. Compared to Sagnac-loop-based detectors, the DO-MZM configuration offers structural simplicity and eliminates complex tuning requirements, though its stability remains susceptible to DO-MZM bias point drift.

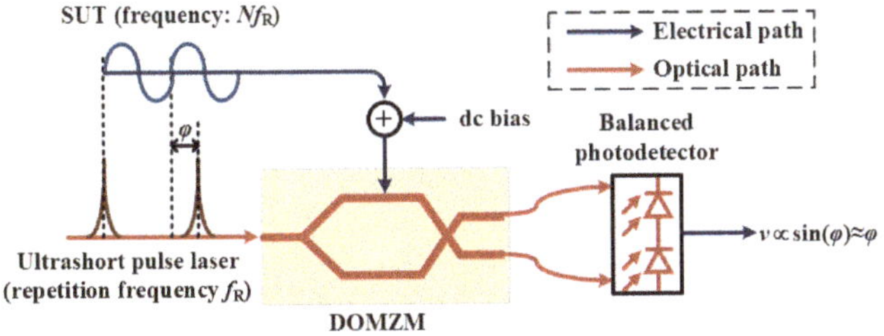

Figure 7. MWP phase detector based on DO-MZM. DO-MZM: Dual-output Mach–Zehnder modulator.

The MWP phase noise measurement methods utilizing ultra-short optical pulse references exhibits exceptionally high sensitivity. However, this approach still faces two primary technical challenges: first, the measurable signal frequency must strictly maintain an integer–multiple relationship with the optical pulse repetition rate; second, the system requires a complex synchronization calibration mechanism to achieve precise matching between optical pulses and the zero-crossing points of the measured signal. To address these limitations, Jingzhan Shi et al. propose an under-sampling phase detection-based PNA [24], shown as Figure 8. This system employs low-jitter, low-repetition-rate pulse sequences as sampling clocks to under-sample the microwave SUT, then applies a phase detection algorithm to extract time-varying phase information from the sampled data. The core concept of the under-sampling phase detection algorithm involves transforming the complex microwave signal phase estimation problem into in-phase and quadrature (I/Q) component estimation. The least squares method is subsequently applied to estimate these components from the sampled data. This under-sampling approach eliminates the requirement for the measured signal frequency to be an integer multiple of the sampling pulse repetition rate, while also removing the need for precise alignment between sampling pulses and signal zero-crossing points. These advancements expand the measurement bandwidth and simplify system architecture. Furthermore, Jingzhan Shi et al. implement a MWP under-sampling phase detection system using ultra-short optical pulses. By leveraging their ultra-low timing jitter characteristics, the system achieves enhanced sampling precision to improve phase noise measurement sensitivity. Simultaneously, the inherent broadband capabilities of MWP sampling systems enable significant expansion of phase noise measurement bandwidth.

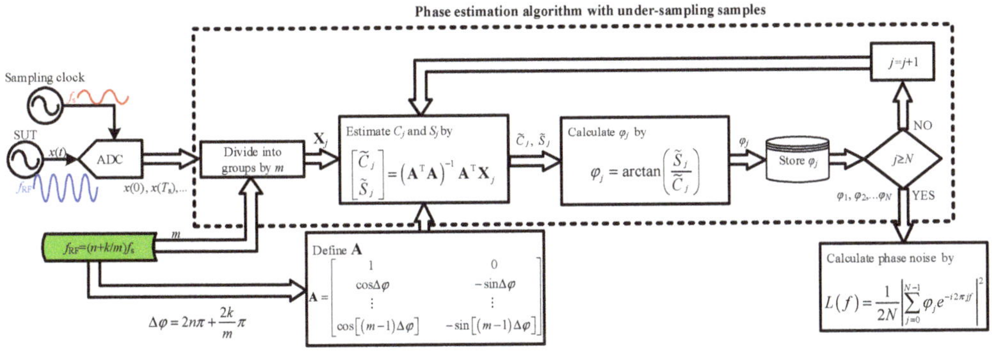

Figure 8. PNA based on phase estimation algorithm with under-sampling samples. ADC: Analog-to-digital converter.

3.2. MWP PNA with OEO

The OEO is a microwave oscillator that utilizes optical fiber as its energy storage component. Generally, the longer the optical fiber, the lower the phase noise of the generated microwave signal [25]. OEOs with ultra-low phase noise characteristics can serve as reference signal sources in phase detection methods, thereby enhancing the sensitivity of phase noise measurements. Zhangyuan Chen et al. proposed a high-sensitivity PNA based on an OEO and a direct digital synthesizer (DDS) [26], with a schematic diagram shown in Figure 9. In this system, the 10 GHz microwave signal generated by the OEO exhibits a phase noise of −140 dBc/Hz @ 10 kHz. By adjusting the DDS output frequency and mixing it with the OEO's 10 GHz signal, the reference source frequency becomes tunable between 9–11 GHz. Consequently, this PNA achieves a measurement bandwidth of 9–11 GHz with a sensitivity of −140 dBc/Hz @ 10 kHz. While the OEO-based MWP phase detection method enables high measurement sensitivity, the limited tunability of the OEO's resonant frequency restricts the phase noise measurement bandwidth.

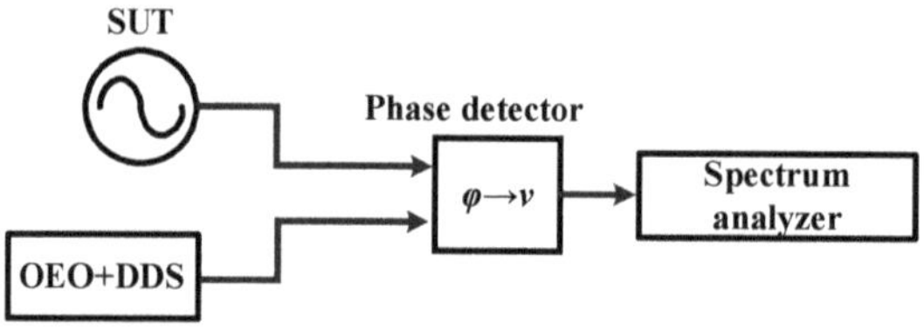

Figure 9. MWP phase-detection-based PNA using OEO as reference. OEO: Optoelectronic oscillator, DDS: Direct digital synthesizer.

4. MWP Frequency-Discrimination-Based PNAs

MWP frequency-discrimination-based PNAs can be categorized into two distinct classes based on their implementation methodologies. The first class involves photonic-substitution-type PNA, where conventional microwave components in the frequency discriminator are systematically replaced with functionally equivalent MWP modules. This architectural substitution primarily aims to enhance both measurement sensitivity and operational bandwidth [8]. The second class constitutes the MWP quadrature-frequency-discrimination-based PNA, which innovatively integrates digital signal processing (DSP) techniques with MWP components. This hybrid approach provides dual advantages: (1) the inherent flexibility of DSP significantly simplifies calibration procedure; (2) photonic-assisted signal conditioning enables simultaneous enhancement of measurement sensitivity and bandwidth.

4.1. Photonic-Substitution-Type PNAs

In the frequency discrimination methodology, the delay line length critically determines the system's sensitivity. However, it is noteworthy that sensitivity at different frequency offsets follow distinct patterns in relation to the delay length. Let $H(f_m, \tau) = 10\lg[\sin^2(\pi f_m \tau)]$. According to Equation (3), the larger $H(f_m, \tau)$ is, the better the sensitivity is. Obviously, the magnitude of $H(f_m, \tau)$ is jointly affected by the delay τ and the frequency offset f_m. In order to analyze the influence of the delay τ on the measurement sensitivity at different frequency offsets, the contour plot of $H(f_m, \tau)$ as shown in Figure 10a is drawn. The abscissa of the contour plot is the frequency offset f_m, ranging from 100 Hz to 100 kHz, and the ordinate is the delay τ, ranging from 1 μs to 100 μs. It can be seen that the influence law of the phase noise measurement sensitivity by the delay τ is different at different frequency offsets:

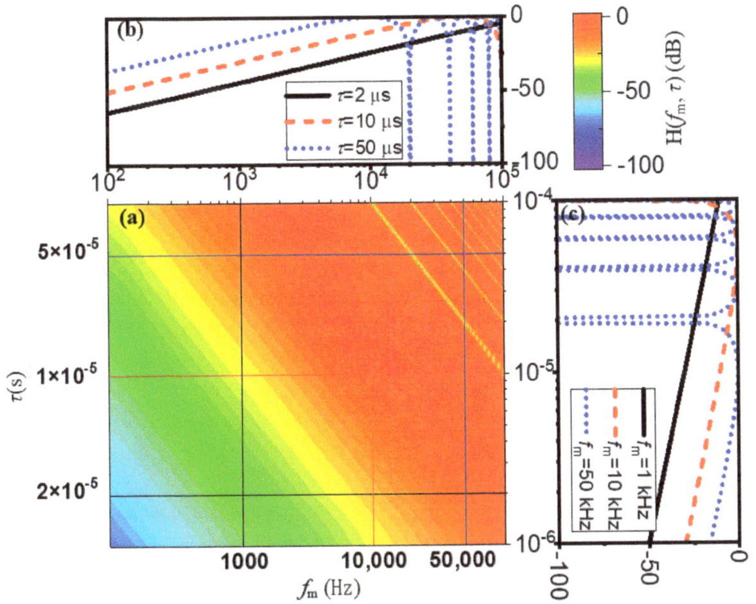

Figure 10. Influence of delay length on the phase noise measurement sensitivity at different frequency offsets for frequency-discrimination-based PNA. (**a**) $H(f_m, \tau)$; (**b**) $H(f_m, 2\ \mu s)$, $H(f_m, 10\ \mu s)$, $H(f_m, 50\ \mu s)$; (**c**) $H(1\ kHz, \tau)$, $H(10\ kHz, \tau)$, $H(50\ kHz, \tau)$.

Indicated by Figure 10b, with fixed delay τ, when frequency offset $f_m \leq 1/2\tau$, measurement sensitivity improves as f_m increases. Within the $f_m \leq 1/2\tau$ range, phase noise measurement sensitivity enhances with longer τ. As shown in Figure 10c, for a specific f_m, when $\tau \leq 1/2f_m$, larger τ yields higher sensitivity at that f_m. Notably, measurement sensitivity degrades near $f_m = k/\tau$ (k = 0,1,2,...), particularly in low-frequency regions (e.g., near 0 Hz). Optimal τ selection depends on target frequency range: for upper frequency bound f_{mu}, one must ensure $\tau \leq 1/f_{mu}$. For specific f_{m0}, optimal $\tau = 1/2f_{m0}$ makes sensitivity at f_{m0} best.

The analytical results demonstrate that enhancing the low-frequency-offset sensitivity in frequency-discrimination-based PNA systems necessitates extended delay lines. For instance, achieving optimal phase noise measurement sensitivity at 10 kHz frequency offset requires a delay duration of 50 μs. However, the implementation of extended microwave transmission lines is inherently constrained by their limited bandwidth, where increased length inevitably induces greater signal attenuation. To address these dual challenges of bandwidth limitation and excessive attenuation in conventional microwave delay systems, E. Rubiola et al. proposed an innovative substitution of coaxial cables with optical fibers for delay generation [27]. Compared with traditional metallic transmission lines, fiber-optic solutions exhibit threefold advantages: broader operational bandwidth (>100 GHz), lower transmission loss (<0.2 dB/km), and significantly reduced mass density. Figure 11 illustrates the microwave PNA architecture employing fiber-optic delay lines. Experimental characterization reveals that with a 2 km fiber delay line configuration, the system achieves remarkable sensitivity of −135 dBc/Hz @ 10 kHz for a 10-GHz carrier, surpassing conventional coaxial implementations by larger than 20 dB.

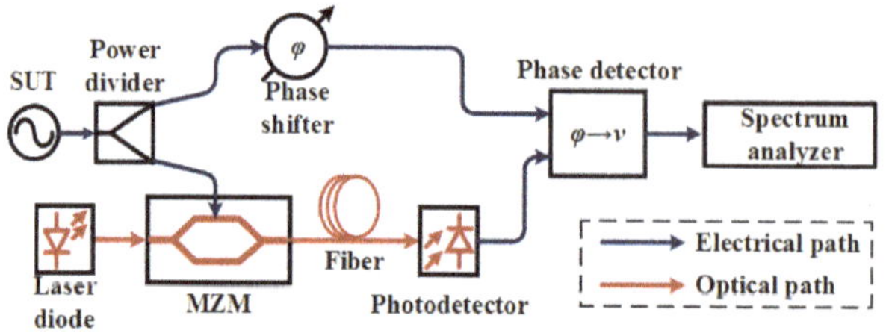

Figure 11. MWP frequency-discrimination-based PNA with photonic delay line. MZM: Mach-Zehnder modulator.

Using optical fiber delay instead of electrical delay can significantly improve the sensitivity of the frequency discrimination method in phase noise measurement. However, since the phase noise measurement sensitivity of the frequency discrimination method drops sharply at frequency offset $f_m = 1/\tau$, the measurable frequency offset range will decrease as the delay τ increases. To solve this problem, B. Onillon et al. proposed a frequency-discrimination-based PNA based on variable delay. The core idea of this method is as follows: when measuring the phase noise at high frequency offsets, a shorter delay is used; when measuring the phase noise at low-frequency offsets, it is switched to a longer delay. This method effectively balances the requirements of measurement sensitivity and the measurable frequency offset range [28].

The introduction of optical fiber delay lines has successfully broken through the limitations of electrical delay lines in terms of sensitivity, while also broadening the phase noise measurement bandwidth to a certain extent. However, the PNAs based on optical fiber delay lines are still limited by other electrical components such as phase shifters and phase detectors. Shilong Pan et al. proposed a PNA based on MWP phase shifter [29,30], aiming to break through the limitations of electrical phase shifter on the phase noise measurement bandwidth. However, the above-mentioned system has limitations when measuring signal sources with a carrier frequency lower than 5 GHz. This is because when the frequency of the SUT is low, the limited roll-off factor of the optical bandpass filter makes it impossible to effectively filter out one of the first-order sidebands while selecting the optical carrier. To address this issue, Ninghua Zhu et al. proposed using up-conversion technology to transform low-frequency signals to higher frequencies [31]. The up-conversion scheme employs a MWP phase shifter based on a Dual-drive Mach-Zehnder modulator (DMZM). By adjusting the DC bias voltage of the DMZM, continuous phase adjustment can be achieved.

A phase detector is generally composed of a mixer and a low-pass filter, where the electrical mixer is the main limiting factor for the working bandwidth of the phase detector. To break through the limitation of electrical mixers on the phase noise measurement bandwidth, Shilong Pan et al. proposed a PNA based on MWP mixing [32]. In this system, the mixer is realized based on a cascaded phase modulator and MZM, enabling the system to achieve a measurement bandwidth of 5–40 GHz. To simultaneously address the limitations of electrical phase shifters and electrical mixers on the measurement bandwidth of the frequency discrimination method, Fangzheng Zhang et al. also proposed a PNA based on all-optical technology [33]. Pei Zhou et al. proposed a microwave PNA based on photonic delay-matched frequency translation [34]. By using a MWP mixer and a local oscillator signal, the SUT is down-converted to a lower intermediate frequency signal, which effectively breaks through the limitation of electrical devices on the working bandwidth and expands the PNA's operating bandwidth. Additionally, by matching the time delays of the two down-conversion branches, the residual phase noise introduced by the local oscillator signal can be effectively suppressed, allowing high measurement sensitivity to be obtained without the need for a high-performance reference source.

To further enhance the sensitivity of microwave phase noise measurement, P. Salzenstein et al. constructed a dual-channel cross-correlation PNA based on the fiber-delay-based frequency discriminator [35–38]. The sensitivity achievable by the dual-channel cross-correlation PNA is related to the number of cross-correlation times. The more cross-correlation times, the higher the sensitivity, but the measurement time also increases accordingly. When the fiber length is 2 km, the carrier frequency is 10 GHz, and 200 cross-correlations are performed, the noise floor of the system at a frequency offset of 10 kHz is −160 dBc/Hz, and the measurement time is 20 min. When 500 cross-correlations are performed, the noise floor at a 10 kHz frequency offset is −165 dBc/Hz, with a measurement time of 45 min.

To improve the phase noise measurement sensitivity without increasing the measurement time, N. Kuse et al. from IMRA America proposed a phase noise measurement method based on an optical frequency comb [39], of which the principle schematic is shown in Figure 12. When the microwave SUT modulates the optical carrier to generate an optical frequency comb, the phase jitter of the nth-order comb is the phase jitter of the optical carrier plus n times the phase jitter of the microwave SUT. The frequency discrimination method is used for the nth-order comb and the -nth-order comb, respectively, to detect the phase difference before and after delay. Then, the phase difference of the nth-order comb before and after delay is subtracted by that of the -nth-order comb before and after delay, so as to remove the phase jitter of the optical carrier and amplify the phase jitter of the microwave SUT by $2n$ times. Correspondingly, the phase noise measurement sensitivity of the system is improved by $(2n)^2$ times.

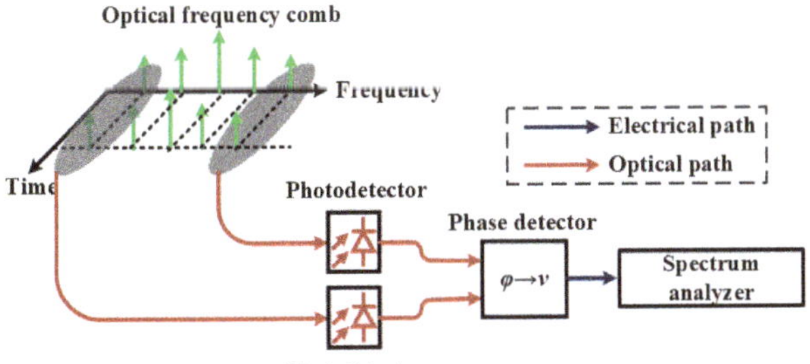

Figure 12. MWP frequency-discrimination-based PNA based on optical frequency comb.

4.2. MWP Quadrature-Frequency-Discrimination-Based PNAs

In a frequency-discriminator-based PNA, the phase detector is typically comprised of a mixer, a phase shifter, and a low-pass filter. This configuration introduces several challenges:

Complexity in calibrating the phase-to-voltage conversion coefficient: The phase-to-voltage conversion coefficient k_{pd} of the phase detector is a critical parameter for phase noise measurement. However, its value is directly influenced by the amplitude of the microwave SUT and the conversion coefficient of the mixer. Given that the mixer's conversion coefficient varies dynamically with the input signal frequency, k_{pd} is affected by both the signal power and frequency. In practical measurements, the power and frequency of the SUT are often unknown and vary each time, necessitating recalibration of k_{pd} before each phase noise measurement. This calibration process is cumbersome and significantly impacts measurement efficiency.

Interference from amplitude noise on measurement accuracy: The phase-to-voltage conversion coefficient k_{pd} of the phase detector is not a constant but a time-varying random quantity that depends on the amplitude jitter of the SUT. However, in current measurement procedures, the calibrated k_{pd} is generally treated as a constant. Consequently, fluctuations in k_{pd} caused by amplitude noise directly introduce measurement errors.

Increased system complexity and error in low-frequency phase noise measurement: To ensure the phase detector operates within its linear region, a phase shifter and feedback loop must be employed to achieve orthogonal bias, thereby increasing the complexity of the system hardware. It is evident that phase jitter is one of the factors affecting the orthogonality of the two signals. When the feedback loop bandwidth covers a specific frequency range, phase jitter at lower frequencies is suppressed by the feedback mechanism, leading to an inability to accurately capture phase noise in this frequency band.

To address the issues caused by the phase detectors, Jingzhan Shi et al. proposed a novel phase noise measurement method—the quadrature frequency discrimination method [40,41]. This method realizes the phase detection function by introducing I/Q mixing and digital signal processing technology, which can effectively avoid the many defects caused by the phase detector in the traditional frequency discrimination method.

The principle of the quadrature frequency discrimination method is shown in Figure 13. Specifically, the microwave SUT is divided into two paths by a power divider: one path is delayed by τ, and then jointly input into the I/Q mixer with the output of the other path of the power divider. After low-pass filtering processing of the output of the I/Q mixer, two orthogonal zero-intermediate frequency signals can be obtained, which are denoted as $v_I(t)$ and $v_Q(t)$, respectively. Subsequently, an analog-to-digital converter (ADC) is used to collect $v_I(t)$ and $v_Q(t)$ and convert them into digital signals for subsequent digital signal processing. The phase of the zero-intermediate frequency signals $\varphi(t) - \varphi(t - \tau) + 2\pi f_c\tau$ (set as $\psi(t)$) can be obtained through

$$\psi(t) = \arctan\left(\frac{v_Q(t)}{v_I(t)}\right) \tag{6}$$

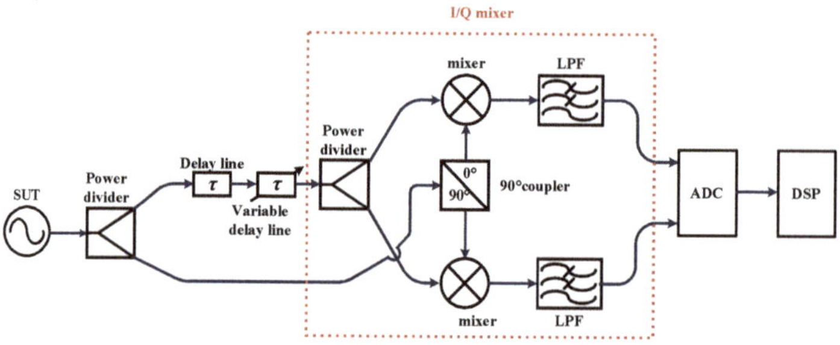

Figure 13. Schematic diagram of quadrature frequency discrimination method. ADC: analog-to-digital converter, DSP: digital signal processing.

Considering that $\psi(t)$ is composed of two terms, $\varphi(t) - \varphi(t - \tau)$ and $2\pi f_c\tau$, and both f_c and τ are constants, the PSD of $\psi(t)$ at non-zero frequencies is identical to that of $\varphi(t) - \varphi(t - \tau)$. According to Equation (3), the phase noise of the microwave SUT can be obtained through the following digital processing:

$$L(f_m) = \frac{[\varphi(t) - \varphi(t-\tau)]_{PSD}}{8\sin^2(\pi f_m \tau)} = \frac{S_\psi(f_m)}{8\sin^2(\pi f_m \tau)}, \quad f_m \neq 0 \tag{7}$$

where $S_\psi(f_m)$ represents the PSD of $\psi(t)$.

In summary, the quadrature frequency discrimination method does not require consideration of the phase-voltage conversion coefficient k_{pd} of the phase detector, thus eliminating the disadvantages of complex calibration and susceptibility to amplitude noise in the traditional frequency discrimination method. Additionally, the quadrature frequency discrimination method does not need a phase shifter or feedback loop to adjust the operating point of the phase detector, thereby overcoming the drawbacks of complex structure caused by the phase shifter and feedback loop in the frequency discrimination method, as well as the impact on phase noise measurement at low-frequency offsets.

It should be noted that for the quadrature frequency discrimination method, I/Q mixing imbalance will affect the accuracy of phase noise measurement. Therefore, it is necessary to compensate for the I/Q mixing imbalance, and the compensation formula can be expressed as

$$\psi(t) = \arcsin\left(\frac{v_Q(t) - k_A v_I(t)\sin(\Delta\psi)}{k_A\cos(\Delta\psi)}\right) \tag{8}$$

It can be seen from Equation (8) that the compensation for I/Q mixing imbalance depends on a precondition: the I/Q mixing amplitude imbalance (k_A) and phase imbalance ($\Delta\psi$) are known. Therefore, in order to accurately compensate for the I/Q mixing imbalance, it is necessary to accurately measure the imbalance of I/Q mixing. Jingzhan Shi et al. proposed an I/Q mixing imbalance measurement method based on adjustable delay. If the delay is changed at a uniform speed v_τ, that is, the delay $\tau = \tau_0 + v_\tau t$, where τ_0 is the fixed delay provided by the delay line and satisfies $\tau_0 \gg v_\tau t$, the trajectory curves of $v_I(t)$ and $v_Q(t)$ shown in Figure 14 can be obtained. Both curves of $v_I(t)$ and $v_Q(t)$ are trigonometric functions with the same period $T = 1/2\pi f_c v_\tau$, and their amplitudes are A and $k_A A$, respectively. The phase difference between the two curves is $\pi/2 - \Delta\psi$. Then, the I/Q mixing amplitude imbalance k_A and phase imbalance $\Delta\psi$ can be derived through the following equation:

$$\begin{aligned} k_A &= \frac{k_A A}{A} = \frac{\mathrm{Amp}(v_Q(t))}{\mathrm{Amp}(v_I(t))} = \frac{\mathrm{Max}_T(v_Q(t))}{\mathrm{Max}_T(v_I(t))} \\ \Delta\psi &= \frac{\pi}{2} - \frac{2\pi\Delta t}{T} \end{aligned} \tag{9}$$

where Amp($y(t)$) represents the amplitude of the function $y(t)$, and $\mathrm{Max}_T(y(t))$ represents the maximum value of the function $y(t)$ within one period T. For the trigonometric function $y(t)$, Amp($y(t)$) = $\mathrm{Max}_T(y(t))$. Δt is the minimum interval between the moments when $v_I(t)$ and $v_Q(t)$ reach their maximum values, as shown in Figure 14.

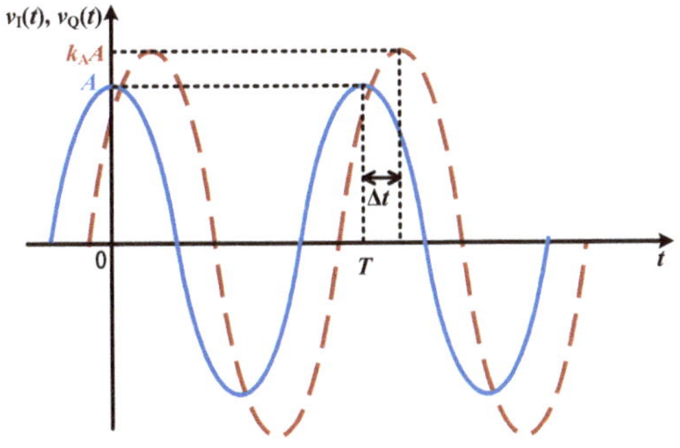

Figure 14. Principle of I/Q mixing imbalance measurement with variable time delay.

As the core component of the quadrature-frequency-discrimination-based PNA, the I/Q mixer is a key factor restricting the measurement bandwidth. In view of this, Zhang Fangzheng et al. proposed a PNA based on optical polarization multiplexing I/Q mixing [42], of which the system structure is shown in Figure 15. This system was experimentally demonstrated with a phase noise measurement bandwidth of 5–35 GHz. To eliminate the dc (direct-current) interference introduced in the MWP I/Q mixing process, they proposed a 180° bridge replacement method to measure dc interference and achieved the removal of dc interference in the digital domain.

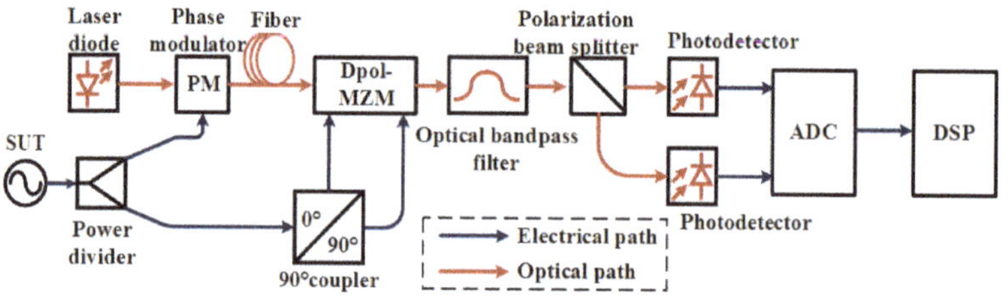

Figure 15. Structural diagram of the quadrature frequency discrimination PNA based on optical polarization multiplexing I/Q mixing. PM: Phase modulator, Dpol-MZM: Dual-polarization MZM.

To break through the limitation of the 90° coupler on the phase noise measurement bandwidth in the quadrature-frequency-discrimination-based PNA based on optical polarization multiplexing I/Q mixing, Shi Jingzhan et al. proposed a PNA based on all-optical I/Q mixing as shown in Figure 16 [43]. This system extends the phase noise measurement bandwidth to 5–50 GHz. Meanwhile, to solve the problems of dc interference and I/Q mixing imbalance simultaneously, they proposed a synchronous measurement method for dc interference and I/Q mixing imbalance based on adjustable delay, and removed dc interference and I/Q mixing imbalance in the digital domain. Compared with the 180° bridge replacement method, the significant advantage of this method is that it does not require reconstructing the phase noise measurement system and can realize the measurement of dc interference and I/Q mixing imbalance simultaneously.

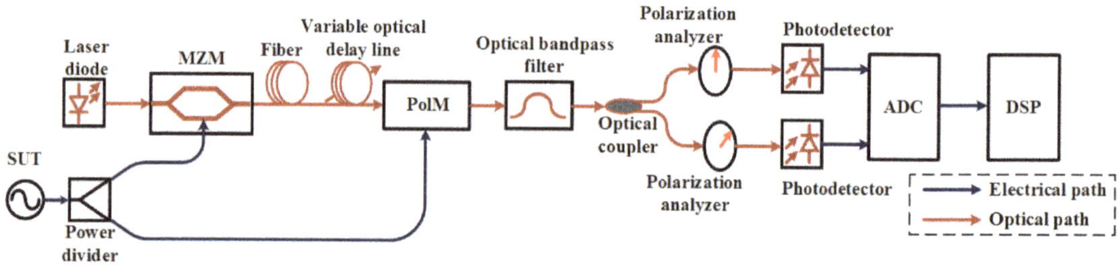

Figure 16. Structural diagram of quadrature frequency discrimination PNA based on all-optical I/Q mixing. PolM: Polarization modulator.

To increase the phase noise measurement bandwidth while improving the phase noise measurement sensitivity, Fangzheng Zhang et al. further proposed a PNA based on optoelectronic balanced detection I/Q mixing [44], as shown in Figure 17. The system is based on the multiplication mechanism of the optical frequency comb for microwave phase noise, which improves the phase noise measurement sensitivity to −146.1 dBc/Hz at a frequency offset of 10 kHz. At the same time, to solve the problem of slow measurement speed caused by the above two dc interference removal methods, they proposed the optoelectronic balanced detection method. Different from the previous two methods that require measuring dc interference first and then removing it in the digital domain, this method directly removes dc interference through optoelectronic balanced detection in the analog domain, which greatly improves the measurement speed.

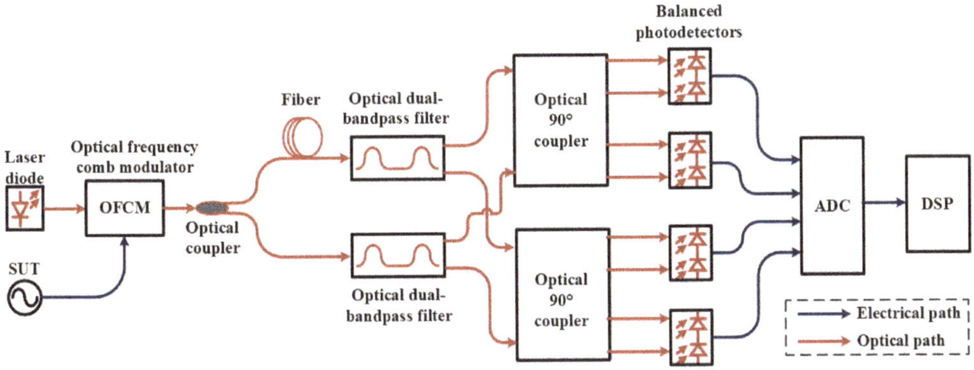

Figure 17. Structural diagram of quadrature frequency discrimination PNA based on photoelectric balanced detection I/Q mixing.

5. Conclusions

In summary, we have systematically reviewed the latest research advancements in microwave photonic phase noise analyzers (MWP PNAs). Two typical MWP phase noise measurement approaches were highlighted: MWP phase detection and MWP frequency discrimination. Table 1 provides a summary of the advantages and limitations of each MWP PNA type, as well as the sensitivity and bandwidth reported in representative studies. Studies have shown that these two approaches can significantly enhance the core performance of microwave signal PNA—not only greatly improving measurement sensitivity but also effectively expanding the measurement bandwidth, providing innovative solutions for high-precision microwave signal analysis.

Table 1. Key performance summary of MWP PNAs: advantages, limitations, sensitivity, and bandwidth.

Types		Advantages	Limitations	Sensitivity	Bandwidth
MWP Phase-Detection-Based PNAs	MWP PNA With Ultra-Short Pulse Lasers	Ultra-high sensitivity	Measurement frequency limitation imposed by laser repetition rate	−167 dBc/Hz @ 10 kHz @ 1 GHz [23] (with dual-channel cross-correlation)	Integer multiples of 1 GHz [23]
	MWP PNA With OEO	High sensitivity	Bandwidth Limitation Caused by Poor Tunability of the OEO	−140 dBc/Hz @ 10 kHz @ 10 GHz [26]	9–11 GHz [26]
MWP Frequency-Discrimination-Based PNAs	Photonic-Substitution-Type PNAs	Wide bandwidth	Complicated calibration; Poor sensitivity at low-frequency offsets	−137 dBc/Hz @ 10 kHz @ 10 GHz [32]	5–40 GHz [32]
	MWP Quadrature-Frequency-Discrimination-Based PNAs	Simplified calibration; Wide bandwidth	Poor sensitivity at low-frequency offsets	−136 dBc/Hz @ 10 kHz @ 10 GHz [43]	5–50 GHz [43]

The current phase noise measurement technology still has several key issues that urgently need to carry out exploration in follow-up research:

(1) Phase noise measurement for complex microwave signals

With the iterative upgrading of electronic information technology, the signal forms adopted by microwave systems such as radar, communication, and electronic warfare show a diversified development trend, typically including linear frequency modulation signals, pulse modulation signals, frequency hopping sequence signals, etc. At the same time, the signal generation capability of modern microwave signal sources has broken through the limitation of single-frequency continuous waves, and can generate more complex waveform combinations. In order to adapt to the development trend of diversified signal forms in microwave source testing, it is urgent to expand the adaptability of phase noise measurement technology to complex signals and achieve precise characterization of phase noise for multiple types of signals [45–47].

(2) Instantaneous phase noise measurement

As a core characteristic parameter of electromagnetic signals, high-precision measurement of phase noise has key application value in fields such as electromagnetic equipment identification and spectrum monitoring [48,49]. However, in complex electromagnetic environments, signals often exhibit burst transient characteristics with extremely short durations (such as nanosecond-level pulse signals), and traditional measurement methods are difficult to capture their instantaneous phase fluctuation characteristics. Therefore, the research on instantaneous phase noise measurement technology is of great practical significance for feature extraction of transient electromagnetic signals and equipment traceability, and related technologies have broad application prospects in scenarios such as radar countermeasures and communication reconnaissance.

Author Contributions: Writing—original draft preparation, J.S.; writing—review and editing, Y.W.; visualization, B.T.; All authors have read and agreed to the published version of the manuscript.

Funding: This research was funded by the National Natural Science Foundation of China, grant number 62101269; the Young Elite Scientists Sponsorship Program by CAST, grant number 2022QNRC001; and the Key Laboratory of Radar Imaging and Microwave Photonics (Nanjing University of Aeronautics and Astronautics), Ministry of Education, grant number NJ20210007.

Conflicts of Interest: The authors declare no conflicts of interest.

Abbreviations

The following abbreviations are used in this manuscript.

PNA	Phase noise analyzer
MWP	Microwave photonic
PSD	Power spectral density
SUT	Signal under test
MZM	Mach-Zehnder modulator
DOMZM	Dual-output Mach-Zehnder modulator
OEO	Optoelectronic oscillator
ADC	Analog-to-digital converter
I/Q	In-phase and quadrature
DDS	Direct digital synthesizer
DSP	Digital signal processing
DMZM	Dual-drive Mach-Zehnder modulator

References

1. Rubiola, E.; Vernotte, F. The companion of Enrico's chart for phase noise and two-sample variances. *IEEE Trans. Microw. Theory Tech.* **2023**, *71*, 2996–3025. [CrossRef]
2. Leeson, D.B. Oscillator phase noise: A 50-year review. *IEEE Trans. Ultrason. Ferroelectr. Freq. Control* **2016**, *63*, 1208–1225. [CrossRef] [PubMed]

3. *IEEE Std 2414™-2020;* IEEE Standard for Jitter and Phase Noise. IEEE Instrumentation & Measurement Society: New York, NY, USA, 2020; pp. 1–42.
4. Ayhan, S.; Scherr, S.; Bhutani, A.; Fischbach, B.; Pauli, M.; Zwick, T. Impact of frequency ramp nonlinearity, phase noise, and SNR on FMCW radar accuracy. *IEEE Trans. Microw. Theory Tech.* **2016**, *64*, 3290–3301. [CrossRef]
5. Rebeyrol, E.; Macabiau, C.; Ries, L.; Issler, J.-L.; Bousquet, M.; Boucheret, M.-L. Phase noise in GNSS transmission/reception system. In Proceedings of the 2006 National Technical Meeting of the Institute of Navigation, Monterey, CA, USA, 18–20 January 2006; pp. 698–708.
6. Mathecken, P.; Riihonen, T.; Werner, S.; Wichman, R. Performance analysis of OFDM with Wiener phase noise and frequency selective fading channel. *IEEE Trans. Commun.* **2011**, *59*, 1321–1331. [CrossRef]
7. Maleki, L. The optoelectronic oscillator. *Nat. Photonics* **2011**, *5*, 728–730. [CrossRef]
8. Pan, S.; Yao, J. Photonics-based broadband microwave measurement. *J. Light. Technol.* **2016**, *35*, 3498–3513. [CrossRef]
9. Zou, X.; Lu, B.; Pan, W.; Yan, L.; Stöhr, A.; Yao, J. Photonics for microwave measurements. *Laser Photonics Rev.* **2016**, *10*, 711–734. [CrossRef]
10. Yao, J. Microwave photonics. *J. Light. Technol.* **2009**, *27*, 314–335. [CrossRef]
11. Seeds, A.J.; Williams, K.J. Microwave photonics. *J. Light. Technol.* **2006**, *24*, 4628–4641. [CrossRef]
12. Rohde, U.L.; Poddar, A.K.; Apte, A.M. Getting its measure: Oscillator phase noise measurement techniques and limitations. *IEEE Microw. Mag.* **2013**, *14*, 73–86. [CrossRef]
13. Lance, A.L.; Seal, W.D.; Labaar, F. Phase noise and AM noise measurements in the frequency domain. *Infrared Millim. Waves* **1984**, *11*, 239–289.
14. Kim, J.; Kaertner, F.X. Attosecond-precision ultrafast photonics. *Laser Photonics Rev.* **2010**, *4*, 432–456. [CrossRef]
15. Kim, J.; Cox, J.A.; Chen, J.; Kärtner, F.X. Drift-free femtosecond timing synchronization of remote optical and microwave sources. *Nat. Photonics* **2008**, *2*, 733–736. [CrossRef]
16. Kim, J.; Song, Y. Ultralow-noise mode-locked fiber lasers and frequency combs: Principles, status, and applications. *Adv. Opt. Photonics* **2016**, *8*, 465–540. [CrossRef]
17. Jung, K.; Kim, J. Subfemtosecond synchronization of microwave oscillators with mode-locked Er-fiber lasers. *Opt. Lett.* **2012**, *37*, 2958–2960. [CrossRef]
18. Jung, K.; Shin, J.; Kim, J. Ultralow phase noise microwave generation from mode-locked Er-fiber lasers with subfemtosecond integrated timing jitter. *IEEE Photonics J.* **2013**, *5*, 5500906. [CrossRef]
19. Kim, J.; Kärtner, F.X.; Perrott, M.H. Femtosecond synchronization of radio frequency signals with optical pulse trains. *Opt. Lett.* **2004**, *29*, 2076–2078. [CrossRef]
20. Lessing, M.; Margolis, H.S.; Brown, C.T.A.; Gill, P.; Marra, G. Suppression of amplitude-to-phase noise conversion in balanced optical-microwave phase detectors. *Opt. Express* **2013**, *21*, 27057–27062. [CrossRef]
21. Kim, J.; Kärtner, F.X.; Ludwig, F. Balanced optical–microwave phase detectors for optoelectronic phase-locked loops. *Opt. Lett.* **2006**, *31*, 3659–3661. [CrossRef]
22. Peng, M.Y.; Kalaydzhyan, A.; Kärtner, F.X. Balanced optical-microwave phase detector for sub-femtosecond optical-RF synchronization. *Opt. Express* **2014**, *22*, 27102–27111. [CrossRef]
23. Endo, M.; Shoji, T.D.; Schibli, T.R. High-sensitivity optical to microwave comparison with dual-output Mach-Zehnder modulators. *Sci. Rep.* **2018**, *8*, 4388. [CrossRef] [PubMed]
24. Shi, J.; Hua, S.; Ge, J.; Zhu, D.; Zhang, F.; Wang, Y. RF Phase Noise Analyzer Based on Phase Estimation Algorithm With Undersampling Samples. *IEEE Trans. Microw. Theory Tech.* **2023**, *72*, 1118–1125. [CrossRef]
25. Maleki, L. The opto-electronic oscillator (OEO): Review and recent progress. In Proceedings of the 2012 European Frequency and Time Forum, Gothenburg, Sweden, 23–27 April 2012; pp. 497–500.
26. Peng, H.; Xu, Y.; Guo, R.; Du, H.; Chen, J.; Chen, Z. High sensitivity microwave phase noise analyzer based on a phase locked optoelectronic oscillator. *Opt. Express* **2019**, *27*, 18910–18927. [CrossRef]
27. Rubiola, E.; Salik, E.; Huang, S.; Yu, N.; Maleki, L. Photonic-delay technique for phase-noise measurement of microwave oscillators. *J. Opt. Soc. Am. B* **2005**, *22*, 987–997. [CrossRef]
28. Onillon, B.; Constant, S.; Liopis, O. Optical links for ultra low phase noise microwave oscillators measurement. In Proceedings of the 2005 IEEE International Frequency Control Symposium and Exposition, Vancouver, BC, Canada, 29–31 August 2005; 6p.
29. Zhu, D.; Zhang, F.; Zhou, P.; Zhu, D.; Pan, S. Wideband phase noise measurement using a multifunctional microwave photonic processor. *IEEE Photonics Technol. Lett.* **2014**, *26*, 2434–2437. [CrossRef]
30. Pan, S.; Zhang, Y. Tunable and wideband microwave photonic phase shifter based on a single-sideband polarization modulator and a polarizer. *Opt. Lett.* **2012**, *37*, 4483–4485. [CrossRef]
31. Wang, W.; Liu, J.G.; Mei, H.; Sun, W.; Zhu, N. Photonic-assisted wideband phase noise analyzer based on optoelectronic hybrid units. *J. Light. Technol.* **2016**, *34*, 3425–3431. [CrossRef]

32. Zhu, D.; Zhang, F.; Zhou, P.; Pan, S. Phase noise measurement of wideband microwave sources based on a microwave photonic frequency down-converter. *Opt. Lett.* **2015**, *40*, 1326–1329. [CrossRef]
33. Zhang, F.; Zhu, D.; Pan, S. Photonic-assisted wideband phase noise measurement of microwave signal sources. *Electron. Lett.* **2015**, *51*, 1272–1274. [CrossRef]
34. Zhou, P.; Jiang, Z.; Tang, Z.; Li, N.; Pan, S. Microwave phase noise analyzer based on photonic delay-matched frequency translation. *IEEE Trans. Microw. Theory Tech.* **2024**, *72*, 5498–5506. [CrossRef]
35. Salik, E.; Yu, N.; Maleki, L.; Rubiola, E. Dual photonic-delay line cross correlation method for phase noise measurement. In Proceedings of the 2004 IEEE International Frequency Control Symposium and Exposition, Montreal, QC, Canada, 23–27 August 2004; pp. 303–306.
36. Eliyahu, D.; Seidel, D.; Maleki, L. Phase noise of a high performance OEO and an ultra low noise floor cross-correlation microwave photonic homodyne system. In Proceedings of the 2008 IEEE International Frequency Control Symposium, Honolulu, HI, USA, 19–21 May 2008; pp. 811–814.
37. Fan, Z.; Qiu, Q.; Su, J.; Zhang, T.; Yang, N. Photonic-delay line cross correlation method based on DWDM for phase noise measurement. *IEEE Photonics J.* **2018**, *10*, 1–9. [CrossRef]
38. Fan, Z.; Qiu, Q.; Su, J.; Zhang, T.; Lin, Y. Phase noise measurement of an optoelectronic oscillator based on the photonic-delay line cross-correlation method. *Opt. Lett.* **2019**, *44*, 1992–1995. [CrossRef]
39. Kuse, N.; Fermann, M. Electro-optic comb based real time ultra-high sensitivity phase noise measurement system for high frequency microwaves. *Sci. Rep.* **2017**, *7*, 2847. [CrossRef]
40. Shi, J.; Zhang, F.; Pan, S. High-sensitivity phase noise measurement of RF sources by photonic-delay line and digital phase demodulation. In Proceedings of the 2017 16th International Conference on Optical Communications and Networks (ICOCN), Wuzhen, China, 7–10 August 2017; pp. 1–3.
41. Shi, J.; Zhang, F.; Pan, S. Phase noise measurement of RF signals by photonic time delay and digital phase demodulation. *IEEE Trans. Microw. Theory Tech.* **2018**, *66*, 4306–4315. [CrossRef]
42. Zhang, F.; Shi, J.; Pan, S. Wideband microwave phase noise measurement based on photonic-assisted I/Q mixing and digital phase demodulation. *Opt. Express* **2017**, *25*, 22760–22768. [CrossRef] [PubMed]
43. Shi, J.; Zhang, F.; Ben, D.; Pan, S. Wideband microwave phase noise analyzer based on an all-optical microwave I/Q mixer. *J. Light. Technol.* **2018**, *36*, 4319–4325. [CrossRef]
44. Zhang, F.; Shi, J.; Zhang, Y.; Ben, D.; Sun, L.; Pan, S. Self-calibrating and high-sensitivity microwave phase noise analyzer applying an optical frequency comb generator and an optical-hybrid-based I/Q detector. *Opt. Lett.* **2018**, *43*, 5029–5032. [CrossRef]
45. Samarasekera, A.C.J.; Feger, R.; Bechter, J.; Stelzer, A. Phase noise measurements in chirped FMCW radar signals. In Proceedings of the 2020 IEEE MTT-S International Conference on Microwaves for Intelligent Mobility (ICMIM), Linz, Austria, 23 November 2020; pp. 1–4.
46. Gerstmair, M.; Melzer, A.; Wagner, M.; Lang, O.; Huemer, M. Phase noise monitoring in cascaded systems for high-resolution automotive radar sensors. *IEEE Trans. Microw. Theory Tech.* **2021**, *69*, 5072–5085. [CrossRef]
47. Birkenhauer, C.; Tschapek, P.; Vossiek, M. Over-the-Air Phase Noise Spectral Density Measurement for FMCW Radar Sensors. *IEEE J. Microw.* **2024**, *4*, 733–741. [CrossRef]
48. Nouri, M.; Mivehchy, M.; Sabahi, M. Target recognition based on phase noise of received signal. *Electron. Lett.* **2017**, *53*, 808–810. [CrossRef]
49. Shi, J.; Zhang, F.; Ben, D.; Pan, S. Photonic-assisted single system for microwave frequency and phase noise measurement. *Chin. Opt. Lett.* **2020**, *18*, 092501. [CrossRef]

photonics MDPI

Review

Advances in and Applications of Microwave Photonics in Radar Systems: A Review

Luka Podbregar, Boštjan Batagelj *, Aljaž Blatnik and Andrej Lavrič

Faculty of Electrical Engineering, University of Ljubljana, Tržaška cesta 25, 1000 Ljubljana, Slovenia; luka.podbregar@fe.uni-lj.si (L.P.); aljaz.blatnik@fe.uni-lj.si (A.B.); andrej.lavric@fe.uni-lj.si (A.L.)
* Correspondence: bostjan.batagelj@fe.uni-lj.si

Abstract: Modern radar systems frequently encounter constraints on bandwidth, transmission speed, and resolution, particularly within complex electromagnetic settings. Microwave photonics (MWP) provides solutions through the integration of photonic elements to improve radar's functionalities. This review paper examines the question of how to improve radar performance by using MWP-based radar components for signal transmission, local oscillator signal generation, radar waveforming, optical beamforming networks, mixing, filtering, co-site interference suppression, real-time Fourier transformation, and analog-to-digital conversion. MWP radar systems achieve wider bandwidths, greater resistance to electromagnetic interference, and reduced phase noise, size, weight, and power consumption. Consequently, the integration of MWP into radar systems has the potential to increase the accuracy of these systems.

Keywords: microwave photonics; microwave photonic radar; optical signal processing

1. Introduction

Conventionally, radar systems have been built solely using electronic technologies. However, these systems face major challenges caused by limited bandwidth, slow transmission rates, and poor resolution, which limit their ability to accurately detect and identify objects in complex electromagnetic environments. To address these restrictions, novel approaches are required, and microwave photonics (MWP) appears to be a promising option. Utilizing the unique features of photonic technologies, MWP can provide significant improvements to radar's performance.

An MWP-based radar system has several advantages over conventional electronic radar systems. It can achieve significantly larger bandwidths, resulting in enhanced resolution. This is useful for a number of applications, including identifying small objects and tracking fast-moving targets. Reduced phase noise is another benefit of using MWP. This means that it can generate more precise reference signals, which can then be used for radar waveform generation or synchronization in distributed radar systems. Additionally, MWP components are immune to electromagnetic interference (EMI). This is crucial for radar systems employed in areas with significant EMI, such as military or industrial settings. Finally, using MWP can result in radar systems that are smaller, lighter, and more power-efficient. This is crucial for numerous applications, including airborne or space-based radar systems.

This review paper examines essential MWP technologies relevant to radar applications, which includes signal transmission, local oscillator (LO) signal generation, radar waveforming, optical beamforming networks, mixing, filtering, co-site interference suppression, real-time Fourier transformation (RTFT), and analog-to-digital conversion (ADC).

The paper seeks to highlight the essential role of MWP in addressing the constraints of conventional radar systems and enabling next-generation radar systems.

2. Microwave Photonics in Radar Systems

MWP is a multidisciplinary research field that merges two distinct areas: microwave or radio-frequency (RF) engineering and optoelectronics [1–4]. MWP offers several key advantages that are crucial to advancing new solutions in the generation, transmission, and processing of RF signals. These advantages include a wide and flexible bandwidth, low-phase noise components for signal generation and detection, immunity to EMI, and the utilization of efficient, scalable optical devices such as optical filters and lasers [5–8]. To take advantage of the photonic properties, broadband electrical-to-optical (EO) and optical-to-electrical (OE) conversions are introduced into radar systems. Figure 1 illustrates the architecture of a photonic radar system, where the RF signal is generated, processed, and transmitted to the antenna's location in the optical domain. OE conversion occurs at the antenna location, which allows the signal to be transmitted by the antenna. After receiving the scattered signal, it is converted back to the optical domain via EO conversion for further processing and transmission in the optical domain.

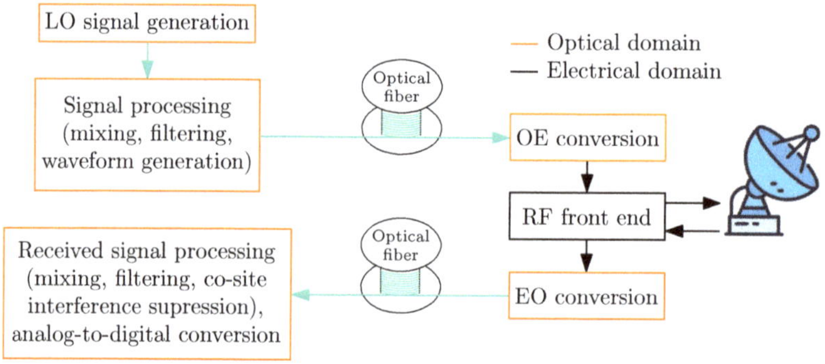

Figure 1. A schematic of a typical photonic radar system. LO: local oscillator; OE: optical-to-electrical; EO: electrical-to-optical.

Radar systems typically incorporate some form of analog signal processing, such as filtering, mixing, phase shifting, and time stretching, among others. These processes are essential for shaping, manipulating, and analyzing the radar signals to enhance performance and improve target detection and identification. Most of the mentioned operations can be easily implemented in the optical domain prior to OE conversion.

MWP-based radar systems provide improved spatial resolution, faster object detection, and reduced system size, weight, and power consumption. The spatial resolution of a radar system is given by

$$\Delta r = \frac{c}{2B}, \tag{1}$$

where c is the speed of light in a vacuum and B is the bandwidth of the radiated radar signal [9]. As shown in Equation (1), the resolution is directly dependent on the signal bandwidth. Increasing the bandwidth of the radiated radar signal improves the spatial resolution, resulting in higher-resolution radar images. However, it is important to note that Equation (1) assumes ideal free-space propagation conditions, without accounting for practical factors such as atmospheric attenuation, scattering, or dispersion, which can become significant at higher frequencies. Therefore, while increasing bandwidth theoretically enhances resolution, the practical limitations imposed by environmental conditions must also be considered in system design and performance evaluations.

Another crucial factor to consider is the signal manipulation capability of radar systems, which is closely related to the relative bandwidth of these systems. The relative bandwidth is defined as the ratio between the radiated signal bandwidth and the central frequency [10]. Modern automotive radar systems typically operate at a central frequency of 77 GHz, with a bandwidth of approximately 4 GHz [11], yielding a relative bandwidth of about 5%.

In contrast, when implementing a radar system with the same 4 GHz bandwidth in the optical domain, where the central frequency is around 193 THz, the relative bandwidth decreases to approximately 0.002%. This substantial reduction indicates that a system considered broadband in the electrical domain becomes a very narrowband system in the optical domain. This characteristic offers greater flexibility for the design of radar systems and improved signal-manipulation capabilities.

Given that spatial resolution is dependent on both central frequency and bandwidth, radar systems are increasingly shifting toward higher frequency bands, such as the millimeter-wave (mmW, 30–100 GHz) band, the sub-terahertz (0.1–0.3 THz) band, and the terahertz (0.3–10 THz) band [12]. These higher central frequencies enable resolutions with sub-millimeter precision [13,14]. However, transmitting these high-frequency signals can be challenging and costly, requiring the use of optimal transmission media.

2.1. *Signal Transmission*

Signal transmission is widely utilized in radar systems, particularly in distributed radar networks, phased-array systems, and those requiring remote signal processing. The most commonly used media for signal distribution in radar systems are coaxial cables, waveguides, and optical fibers. Although coaxial cables are suitable for short distances and low frequencies, they become impractical for longer distances and higher frequencies due to high losses [15] and their cost. Nevertheless, their widespread availability makes them popular in modern radar applications. While waveguides are advantageous for high-power radar applications, they lack flexibility due to their rigidity. They are also costly for high-frequency applications and highly sensitive to mechanical vibrations, making them suitable only for permanent installations [16]. However, filling them with dielectrics can make them cheaper and more flexible. Polymer microwave fiber (PMF), a type of dielectric waveguide, has emerged as a promising technology for transferring microwave signals. PMFs are low-cost, highly temperature insensitive, lightweight, and mechanically robust [17]. Signals can be distributed with a PMF up to 25 m [18,19]. Optical fibers offer a viable alternative, offering key advantages such as lower losses, higher bandwidth, and immunity to EMI. This allows for the distribution of the signal over longer distances and the implementation of long optical delay lines, which can be used as high-quality optical storage elements in setups such as optoelectronic oscillators (OEOs). However, temperature fluctuations and mechanical vibrations cause delay fluctuations in the transmission of the signal through optical fibers, which can affect the phase of the received signal [20]. These delay fluctuations can be mitigated using various compensation methods designed to counteract these issues [21–25].

Compensation methods can be categorized into two groups: methods in which the phase of the transmitted signal is actively [22] or passively [23] adjusted and methods in which the phase is controlled by altering the physical properties of the transmission medium [24]. These properties include the fiber length and its refractive index. Additionally, compensation can be performed within the optical link by adjusting the laser wavelength [25]. Such methods enable phase-stable signal transmission, which is essential for achieving high spatial resolution in coherent distributed radar systems [26]. However, these techniques can be complex, costly to implement, and are not yet widely adopted in modern radar systems.

The use of optical fibers for transmitting RF signals is known as the Radio-over-Fiber (RoF). This technique offers several advantages, including low-attenuation broadband transmission, significantly reduced costs compared to those of other methods, the ability to transmit through multiple independent channels, and immunity from eavesdropping and EMI. Additionally, optical fibers are compact, flexible, and easy to install, making them particularly well suited to confined spaces, such as inside vehicles [27,28].

The transmission of multiple independent signals is essential in complex, modern radar systems. In distributed multiple-input multiple-output (MIMO) radar systems with coherent processing, it is necessary to distribute the LO signal, trigger signals for ADC synchronization, and large amounts of raw data. Independent, high-throughput, bidirectional signal transmission can be achieved using optical fibers with various multiplexing techniques, the most common being wavelength-division multiplexing (WDM), polarization-division multiplexing (PDM), space-division multiplexing (SDM), and mode-division multiplexing (MDM).

- **WDM** enables the simultaneous transmission of multiple optical signals by assigning each signal a unique wavelength, allowing them to travel through a single optical fiber without interference [29]. This technique takes advantage of the large bandwidth available in optical fibers to significantly increase data capacity. Figure 2 illustrates a typical WDM system, which employs a multiplexer on the transmitter side to combine the input signals and a demultiplexer on the receiver side to separate them. Widely adopted in telecommunications, WDM improves fiber data throughput without requiring any additional physical infrastructure.
- **PDM** is a technique that transmits two independent data channels on a single wavelength using orthogonal polarization states [30–32]. Although multiplexing two optical beams with orthogonal polarizations in an optical fiber is relatively straightforward, demultiplexing them can be challenging. This is due to the constant variation in the polarization states within single-mode fibers (SMFs), even though the polarizations remain orthogonal. Furthermore, polarization-maintaining fibers (PMFs) are limited to supporting a single polarization state, making them less suitable for PDM [33].
- **SDM** employs multiple, spatially separate channels within a single optical fiber to enhance data transmission [34]. In the literature, different configurations that use multicore fiber (MCF) or multimode fiber (MMF) have been proposed as promising candidates for next-generation multiplexing technologies in optical fiber communications [35].
- **MDM** is a multiplexing technique that uses different guided modes in a MMF for different transmission channels [36–39]. Typically, the MMFs used in such applications support a relatively small number of guided modes and are therefore referred to as few-mode fibers.

Combinations of various multiplexing techniques have also been extensively studied to maximize transmission performance. For example, in [40], hundreds of WDM channels were transmitted through more than 100 spatial channels, achieving an impressive transmission rate of 10.66 Pb/s. Similarly, different studies have demonstrated the use of combined techniques, such as PDM and WDM [41], MDM and WDM [42], and other multiplexing methods.

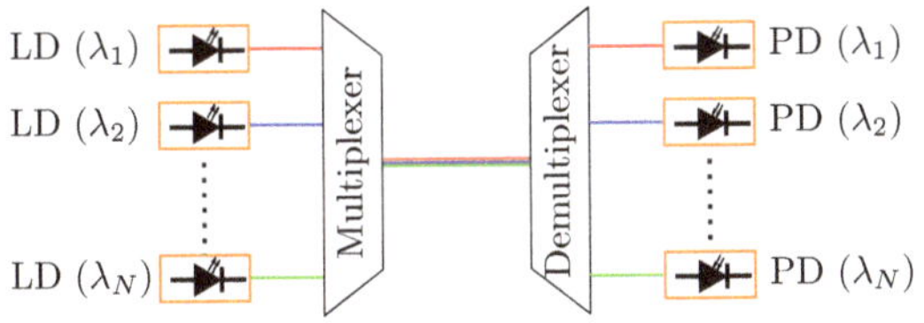

Figure 2. Schematic of typical WDM optical transmission system. LD: laser diode; PD: photo diode.

2.2. Local Oscillator Generation

LO signal generation is a crucial component of radar systems, providing reference signals for radar signal generation, synchronization, and timing for ADCs, digital-to-analog converters (DACs), and digital signal processing (DSP) modules. To meet the growing performance demands of modern radar systems, photonic techniques for LO generation have been proposed [8,43–57].

An important characteristic of an LO is its short-term frequency stability, also known as the phase noise. The low phase noise of the LO in a radar system is particularly crucial to accurately extracting weak reflections from targets. Another key characteristic is the spectral purity, which can be compromised by harmonics, sub-harmonics, and non-harmonic spurs generated by the LO. Traditionally, high-frequency LO signals are generated by multiplying the output of a low-frequency electrical oscillator, such as an oven-controlled crystal oscillator or a dielectric resonator. However, frequency multiplication in the electrical domain suffers from high phase noise, resulting from the multiplication factor, and a high spur level that require quality filtering to remove. Using photonic methods for LO generation in a radar system can significantly improve its performance, particularly in terms of phase noise, long-term frequency stability, spectral purity, complexity, and reliability. A comparison of various photonic LO-generation methods shows that techniques based on optical frequency multiplication [43,58–61], optical–microwave synchronization [44–46,62], and OEOs [47–57] achieve performances comparable to or better than their electronic counterparts.

The beating of two spectral lines can result in the generation of signals with a very low phase noise. One method is optical–microwave frequency multiplication, which works by using a low-frequency, narrow-band RF signal, generated in the electrical domain, to modulate a high-quality laser source via an MZM [58]. Figure 3 shows a typical optical frequency multiplier, which uses four MZMs to create a dual-band LMF radar signal. By adjusting the DC biases of the electro-optical modulator (EOM) or by incorporating an optical filter, specific sidebands can be selected. Sideband selection enables the generation of signals with frequencies that are two, four, or eight times or even higher multiples of the input RF frequency and is a key advantage of optical–microwave frequency multiplication. However, the quality of the generated signal is directly dependent on the quality of the input RF signal and its phase noise [59]. To improve the phase noise, mode-locked lasers (MLLs) can be used as the source for the spectral lines [60]. However, various imperfections and carrier fluctuations in the photodiode used in this process can increase the phase noise. A solution using an optical–microwave phase detector for optical–microwave synchronization was proposed to overcome the problem mentioned [45,46,62]. It works by converting optical pulses to an electrical signal using a photodiode and then using an optical–microwave phase detector to detect the phase difference between the microwave signal and the optical pulses [44]. Feedback is applied to the MLL or its repetition rate to achieve synchronization. Applying this method to an MLL can result in ultra-low-noise synchronization between the components, allowing for ultra-low-phase noise LO signal generation.

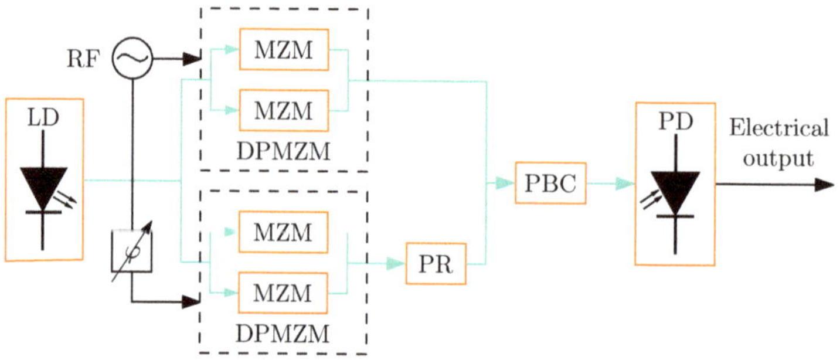

Figure 3. Scheme of signal generation based on optical frequency multiplication using DPMZM. DPMZM: dual parallel Mach–Zehnder modulator; MZM: Mach–Zehnder modulator; PBC: polarization beam combiner; PR: polarization rotator.

A promising solution for low-noise LO generation is an OEO [47–57]. This is a simple and cost-effective microwave photonic system capable of producing RF signals with ultra-low phase noise by utilizing a high-quality optical storage element, such as a long optical fiber delay line or a high-quality optical resonator. A typical single-loop OEO structure can be seen in Figure 4. Continuous-wave (CW) light from a laser source is modulated using an EOM, usually a Mach–Zehnder modulator (MZM), and then passed through a long optical storage element. OE conversion is performed using a photodiode, after which the electrical signal is amplified, filtered, and fed back to the MZM, creating an oscillation loop. The maximum achievable frequency of OEOs is determined by the bandwidth of the optical and electrical components, which can reach up to 100 GHz and beyond [63].

However, OEOs tend to be large and bulky and consume significant power compared to simple microwave generators, due to their complex design [64]. When using a long optical fiber delay line as a high-quality storage element, the temperature sensitivity of the optical fiber must be considered, as it can induce phase fluctuations in the transmitted signal [65,66]. These issues become even more pronounced at higher frequencies, further limiting the operating frequency of OEOs. To mitigate the effects of these temperature fluctuations, methods for stabilizing the temperature of the fiber spool and RF cavity have been proposed [67].

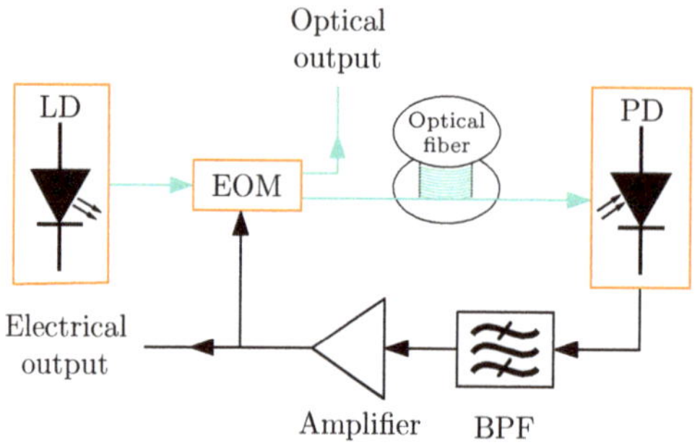

Figure 4. A schematic of a typical single-loop OEO. EOM: electro-optical modulator; BPF: electrical bandpass filter.

2.3. *Radar Waveforming*

Radar systems require specific signal shapes or waveforms that are adapted to desired performance characteristics. In most cases, the shape of the radar signal is adjusted based on the required resolution for measuring distance and speed. Figure 5 shows the typical waveforms used in radar systems. Using more advanced signal shapes, it is possible to maximize the pulse energy while minimizing the peak power, improve the spectrum efficiency, or enable advanced signal processing. The two most common radar waveforms are frequency-modulated continuous-wave (FMCW) signals and pulsed radar signals. FMCW signals are widely used in automotive applications, while pulsed radar signals are more prevalent in military systems.

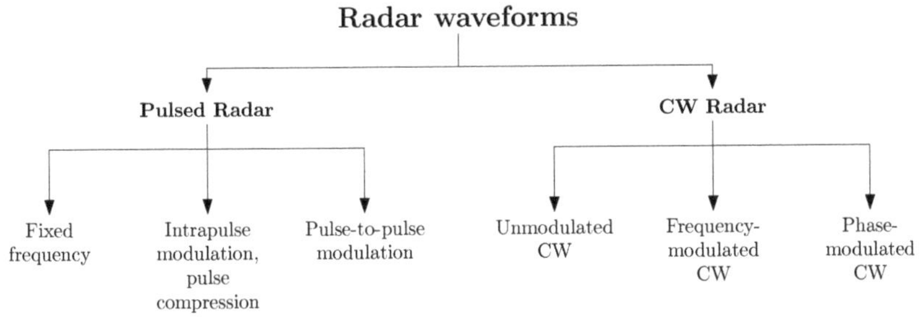

Figure 5. Typical waveforms used in radar systems. CW: continuous wave.

Classical radar systems generate radar signals either in the analog domain using a microwave voltage-controlled oscillator or digitally using a direct signal synthesizer. These methods often face challenges due to a limited bandwidth. To address these limitations, photonic techniques have been proposed for generating specific waveforms. In general, photonic methods for generating microwave radar signals can be categorized into several approaches: spectral shaping and frequency-to-time domain mapping [68–70], external optical injection locking [71–74], photonic multiplication [61,75,76], and photonic DAC [77–80].

Spectral shaping and frequency-to-time mapping works by shaping an ultrashort optical pulse to match the desired waveform type. This method can achieve signal bandwidths of up to 37.4 GHz, but it suffers from a very limited time duration [70]. As a result, it is not suitable for radar systems requiring long-distance detection. Optical injection locking is a technique for synchronizing optical frequency and phase, based on the photon–photon interaction when external light is injected into a laser cavity [81]. The system consists of a primary and secondary laser, where the primary laser is injected into the secondary laser via a circulator or an isolator (to avoid any parasitic reflections back to the primary laser). By optically injecting the secondary laser, it can enter a period-one (P1) oscillation stage, where the laser's output power oscillates periodically with a single, fundamental frequency. This frequency can be controlled by both the injection strength and the frequency detuning between the primary and secondary laser. The technique can be used to generate tunable microwave signals with frequencies that can be much higher than what can be easily achieved with purely electronic oscillators. The tuning is controlled optically, offering broad bandwidth and potential for integration. In [73] a linearly frequency-modulated (LFM) signal with a 12-GHz bandwidth was generated using optical injection locking.

Another promising photonic technique is the use of a photonic DAC to generate radar waveforms. In [77], sawtooth, triangular, parabolic, and rectangular waveforms were generated using a photonic DAC, while in [79], an LFM waveform with a 4 GHz bandwidth was produced. However, most implementations suffer from poor linearity, limited time duration, and limited bandwidth. The limited time duration of the generated waveforms, combined with their constrained bandwidth, affects both the range resolution and the maximum detectable range of pulse radar systems, which can significantly limit their practical applicability in long-range scenarios [82].

2.4. Mixing

Photonic mixers have emerged as an important area of research [6,12,77,83–89]. These are crucial in radar systems for the up-conversion of the generated waveform signal to the RF band on the transmitter side and for the down-conversion back to the IF band on the receiver side, as shown in Figure 6. Mixers produce unwanted mixing products after mixing, which are then usually filtered out. However, this can be hard to carry out if the input signal has a wide bandwidth, which can cause the unwanted mixing products to overlap with the desired mixing products. Electrical mixers are very limited in terms of operational instantaneous bandwidth and their dynamic range. To solve these problems, the use of photonic mixers based on EOMs has been explored for several photonic radar systems [6,12].

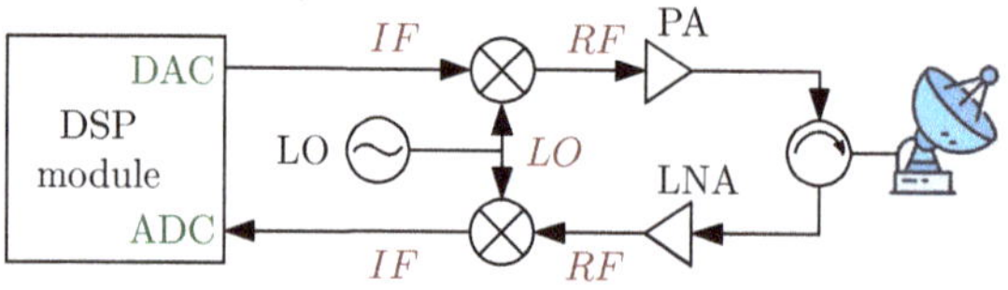

Figure 6. Typical usage of mixers in radar systems. DSP: digital signal processing; DAC: digital-to-analog converter; ADC: analog-to-digital converter; PA: power amplifier; LNA: low-noise amplifier.

Although a photonic mixer is not strictly necessary for photonic radar systems, particularly on the receiver side, it offers key advantages, such as higher bandwidth, lower phase noise, and reduced size [6,12,77]. It also allows for near-infinite isolation between the RF and the LO ports, as well as EMI immunity. The most common method for photonic mixing is using the serial connection of two EOMs, where the RF signal and LO signal are applied in two separate EOMs [84,85].

Another use of mixers is in signal generation, where a high-frequency signal with a wide bandwidth can be realized with photonic mixing, by combining two separate laser sources using a fast photodiode (photomixing). This method enables an extremely wide signal generation range, constrained only by the frequency range of the lasers, potentially spanning several terahertz, and the bandwidth of the photodiode used [87]. The laser sources can also be modulated with certain signals using MZMs, as shown in Figure 7. To minimize the phase noise of the system, the two laser sources should be correlated as much as possible. Techniques such as phase-locked optical loops, optical injection locking, or the use of an external modulator to simultaneously modulate both lasers have been proposed to enhance the correlation [88]. Some techniques for LO generation and radar forming are explained in Sections 2.2 and 2.3.

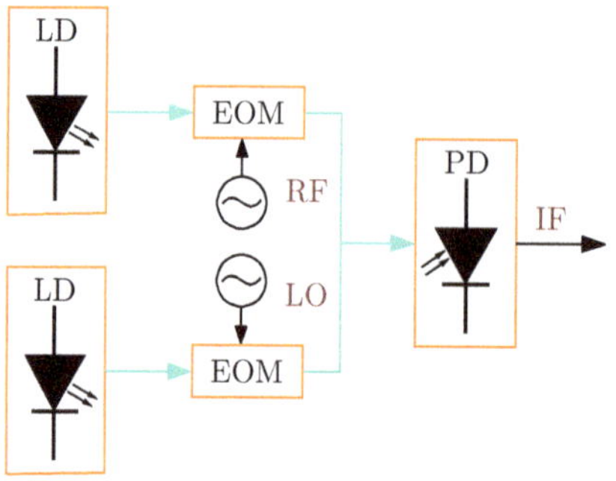

Figure 7. Principle of photomixing of two modulated laser sources with frequencies RF and LO. Result of photomixing is electrical signal with frequency IF.

2.5. Filtering

In modern radar systems, unwanted frequency components, such as crosstalk between the transmitter and receiver or unwanted mixing products, are typically filtered in the digital domain using digital signal processing (DSP). However, DSP is limited by the dynamic range of the ADC used. To overcome these limitations, many radar systems still rely on analog filters for the signal filtering. Although filtering electrical RF signals is a well-established and widely used technique, it becomes more challenging at higher frequencies, especially when tunable narrow-band filters are required.

To address these challenges, several solutions that incorporate optical elements have been proposed for signal filtering in radar systems [90–93]. Photonic filtering is particularly advantageous when photonic components are already integrated into the system, as no additional EO- or OE-conversions are needed. Filtering in the optical domain is a more straightforward process, allowing easy tuning by adjusting the physical properties of the optical filter [7,94,95].

The most used method for implementing MWP filters relies on discrete-time signal processing, where many weighted and delayed samples of the RF signal are generated in the optical domain and subsequently combined upon detection. Finite impulse response (FIR) filters combine a finite number of delayed and weighted replicas, or taps, of the input optical signal at their output, while infinite impulse response (IIR) filters use recirculating cavities to generate an infinite series of weighted and delayed replicas of the input optical signal [90]. Figure 8 shows a general scheme of a discrete-time MWP FIR filter.

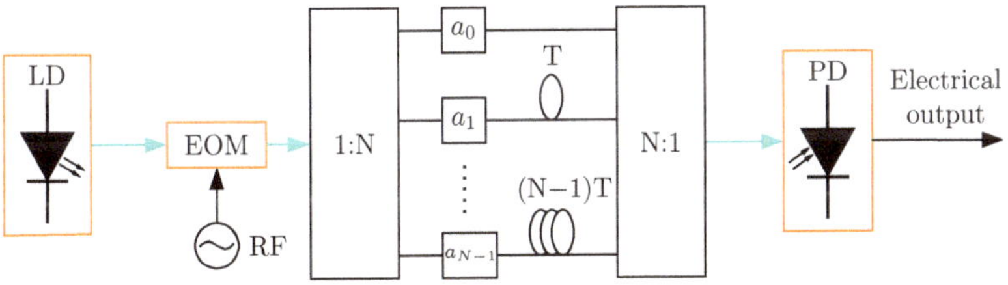

Figure 8. A scheme of a typical discrete-time MWP FIR filter.

2.6. Co-Site Interference Suppression

During a radar system operation, part of the signal emitted by the transmitter may be received as cross-talk at the receiver. This is undesirable because it can saturate the receiver and prevent the detection of weak signals reflected from the target. One straightforward solution is to use pulse radar, which alternates the operation of the transmitter and receiver. In this method, the transmitter emits a signal for a short period while the receiver is disabled. The duration of the pulse must be much shorter than the time it takes for the signal to travel to the target and back, ensuring that the transmitter is off when the receiver collects the reflected signal. However, this approach introduces "dead zones", which are periods when the radar system is unable to detect any targets.

Another common approach is to use advanced radar waveforms, such as FMCWs, where the radar transmits continuously but at varying frequencies. Since reflection from the target is delayed, the instantaneous frequency of the transmitted signal differs from that of the received signal. This difference allows the receiver to distinguish between co-site interference and the reflected signal. In conventional systems, co-site-interference cancellation is typically performed in the electrical domain, but this approach faces limitations at high frequencies and large bandwidths. To overcome these constraints, various optical solutions have been proposed to cancel the co-site interference at higher frequencies and over large bandwidths [91,92,96–98]. Figure 9 shows a classical method for suppressing the co-site interference, i.e., coherent cancellation in the optical domain, where a mitigation signal with the same amplitude but opposite phase to the interference is generated and then combined coherently with the received signal [91,92,97,98].

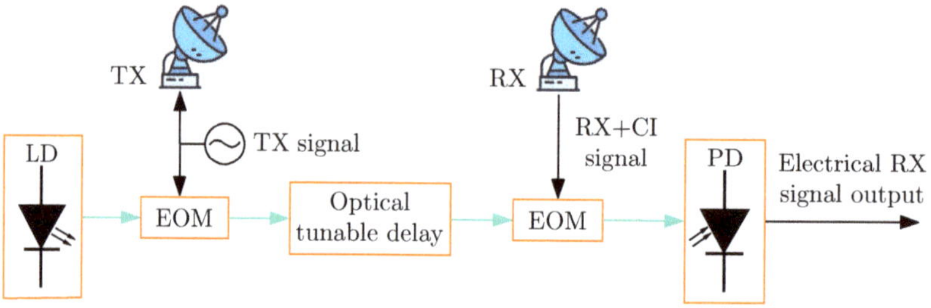

Figure 9. A scheme of a typical method for co-site interference suppression. TX: transmitter; RX: receiver; CI: co-site interference.

2.7. Optical Beamforming Networks

Radar-phased array networks are commonly used in various remote sensing applications. Traditionally, these networks were implemented using phase shifters, but their bandwidth is very limited due to beam squint [99,100]. Beam squint is an effect where the beam angle changes depending on the frequency and is closely related to the signal's bandwidth. The greater the signal bandwidth is, the more pronounced the beam squint becomes. By employing optical beamforming networks, the benefits of microwave photonic links can be integrated into phased-array radar systems [101–103]. This can also solve the problem of beam squint.

The scheme of a typical optical beamforming network is shown in Figure 10. It consists of a laser source modulated by an RF signal, which is then transmitted through a delay element and detected by a photodiode. To achieve an RF frequency response with a controllable linear phase, the phase delay introduced by the delay element must have a linear phase response over the operational optical band [8]. Solutions are based on the adjustment of the length or the dispersive properties of the transmission fiber [103–108], Fourier-domain optical processors [109] and thermally tuned integrated ring resonators [110] have been proposed to introduce tunable delays in fiber links.

While fiber-based delay lines provide high delay-bandwidth products and flexibility, their reliance on long fiber spools introduces practical challenges, such as increased system size, added weight, and mechanical stability issues, especially in airborne and mobile radar applications. To address these challenges, recent research has explored integrated photonic beamforming approaches using compact tunable delay elements and photonic integrated circuits [111,112], which offer a smaller footprint, improved mechanical stability, and lower power consumption.

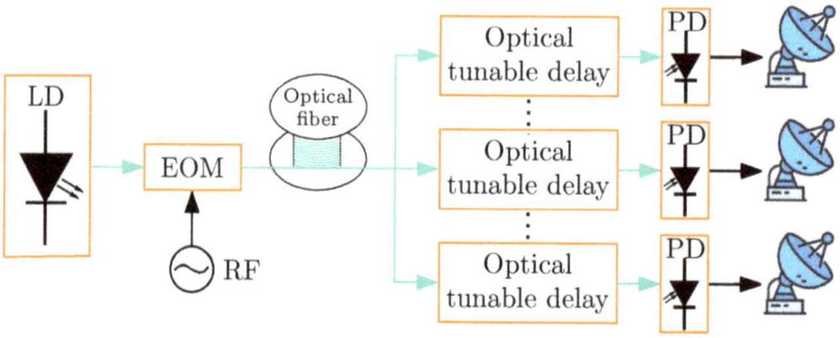

Figure 10. A scheme of a typical optical beamforming network.

2.8. Optical Real-Time Fourier Transformation

Conventional radar uses digital processing to implement spectral analysis via the Fourier transformation, which is crucial for radar signal processing. However, this method is constrained by the quality, speed, and bandwidth of the ADC in the system. To overcome the limitations of digital spectral analysis, the use of photonic elements for optical RTFT has been proposed [113–121].

RTFT is based on the concept of space–time duality. This refers to a parallel between the paraxial diffraction of beams through space and the dispersion of narrow-band pulses through dielectric media in time [114]. The far-field or Fraunhofer diffraction regime represents the simplest spatial propagation geometry that exemplifies this duality. As illustrated in Figure 11a, when light diffracts through a slit, it generates a spatial pattern at a distant point in the far field that resembles the Fourier transformation of the incident beam. For this phenomenon to occur, the distance from the screen has to be

$$z \gg kx^2, \tag{2}$$

where k is a propagation constant and x is the size of the input aperture. A similar phenomenon occurs when a signal propagates through a highly dispersive medium, as shown in Figure 11b. There, the dispersive medium's group delay dispersion D has to be

$$D \gg \tau_w^2, \tag{3}$$

where τ_w^2 is the temporal width of the input waveform [122]. By modulating the optical source, the RTFT of a wanted signal can be realized. The frequency resolution of the dispersion-based RTFT is determined by the dispersion value of the dispersive element; a larger dispersion value results in a higher frequency resolution.

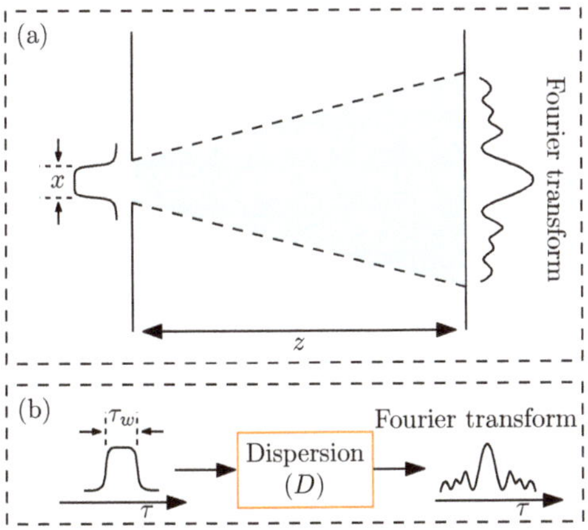

Figure 11. (**a**) A scheme of the far-field or Fraunhofer diffraction of a uniformly illuminated slit. (**b**) The equivalent of the far-field diffraction in the time domain caused by dispersion.

Another method of RTFT involves the use of a time lens, which is a cascade of a dispersive element (such as an optical fiber or fiber grating), a phase modulator, and a dispersive element [123]. Figure 12a illustrates a scheme for time-lens dispersion-based RTFT. A temporal convolution-based RTFT method is shown in Figure 12b. In this method, an ultrashort optical impulse is generated and then stretched by a dispersive element. The pulse is subsequently modulated with an RF signal in MZM and compressed back by another dispersive element. The spectrum of the RF signal can then be measured using a high-sample-rate oscilloscope connected to a photodiode [117,124]. Figure 12c demonstrates the realization of RTFT using a frequency-shifted loop, which has the advantage of not having a limitation on the time window and can therefore be applied to the measurement of infinitely long signals. However, its bandwidth is limited to only tens of MHz, constrained by the frequency shift of the acousto-optic modulator [125].

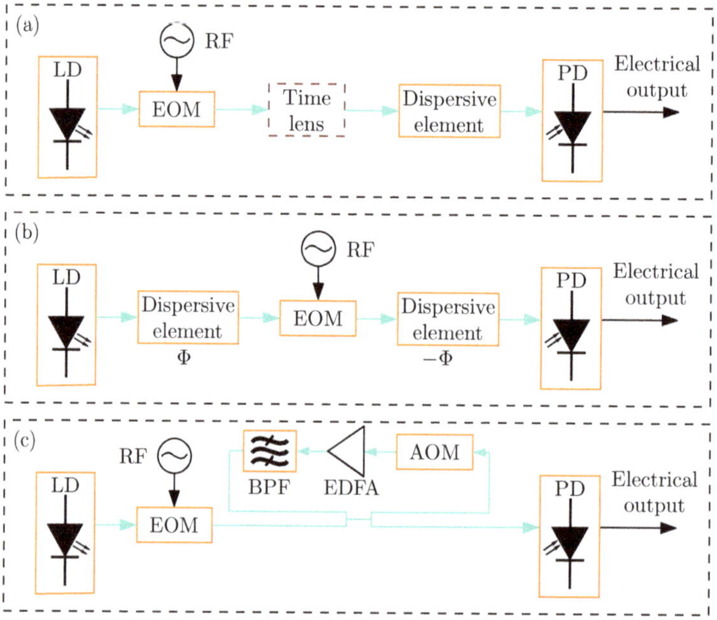

Figure 12. Schemes of typical RTFT methods: (**a**) Dispersion-based RTFT. (**b**) Temporal convolution-based RTFT. (**c**) Frequency-shifted loop-based RTFT [8]. DE: dispersive element; MLL: mode-locked laser; AOM: acousto-optic modulator; EDFA: erbium-doped fiber amplifier.

In recent years, another promising technique for RTFT has emerged using integrated silicon photonics [119–121]. This approach enables the miniaturization of optical spectrometers but introduces performance trade-offs compared to traditional implementations. These include limitations in operating bandwidth, measurement speed, spectral resolution, and dynamic range. The compact size and low power consumption of integrated photonic chips make them a viable solution for RTFT in applications where size, weight, and power are constrained, such as in drones. However, this technology has yet to be tested in radar systems.

2.9. Analog-to-Digital Converters

With the increasing demand for multifunctionality and software-defined operation in modern radar systems, it is widely accepted that as much signal processing as possible should be performed in the digital domain. As a result, high-performance ADCs are essential for integrating radar systems with DSP systems. However, as radar systems migrate to higher frequencies, ADCs require higher sampling rates, wider analog bandwidths, and a higher effective number of bits, which are challenging to achieve with traditional ADCs [8].

Using photonic components with wide bandwidth and high stability, we can significantly enhance the performance of ADCs [126,127]. Two main types of photonic ADCs have been demonstrated in radar systems: photonic sampling ADCs and photonic preprocessing ADCs. Using an MLL, optical pulses in the femtosecond range can be generated [128], functioning similarly to Dirac pulses, which are ideal for sampling analog signals. A schematic of a photonic sampling ADC is shown in Figure 13. In this setup, the MLL generates short pulses that are intensity modulated by an EOM. The modulated optical pulses are then converted to the electrical domain via a photodiode and subsequently sampled by an ADC synchronized with the MLL. This approach enhances the analog bandwidth of the ADC, although it does not directly increase the sampling frequency. However, with minor modifications, it allows for parallelization or the use of multiple ADCs, thereby summing the sample rates of all ADCs. The concept of photonic sampling ADCs can be traced back to [129,130].

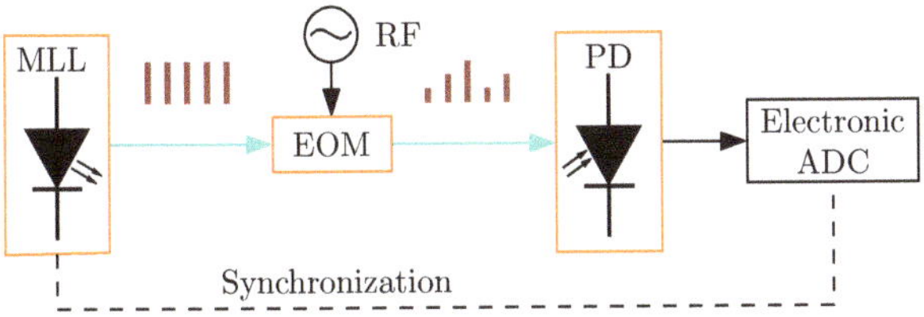

Figure 13. A scheme of a typical photonic sampling ADC.

Photonic preprocessing enables the simpler conversion of signals to the digital domain. A schematic of a photonic preprocessing ADC is shown in Figure 14. In this method, signal preprocessing involves the expansion of the time domain [131]. The short pulse from the MLL is modulated by the analog signal, broadened spectrally, and then converted to the electrical domain via a photodiode. The time-expanded signal is fed into the ADC, which simplifies conversion to the digital domain. However, a drawback of this approach is the slower conversion speed, which limits its application in systems that require rapid processing.

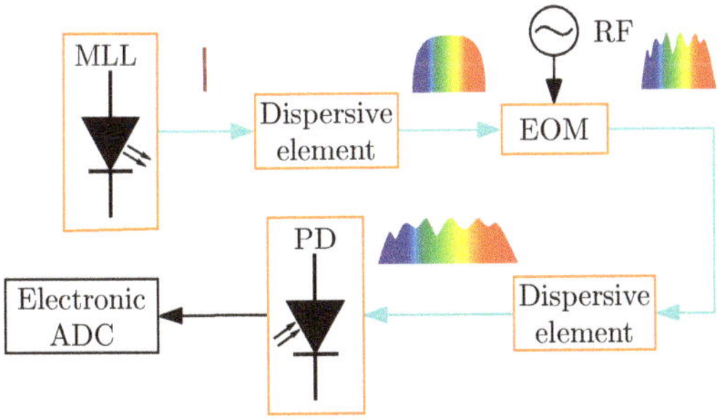

Figure 14. A scheme of a typical photonic preprocessing ADC.

Recently, integrated photonic ADCs have been gaining a lot of attention [132]. Various approaches have been proposed, including schemes based on spectral slicing [133,134], time and wavelength interleaving [80,135], and combinations of wavelength-division multiplexing (WDM) and mode-division multiplexing (MDM) [136]. While some of these techniques show promising results, most have yet to achieve performance levels that are satisfactory for practical radar applications.

3. In-Field Demonstrations of Microwave Photonic Radar

As highlighted in this review paper, many applications of MWP in radar systems have been demonstrated in the literature. However, relatively few have been validated in real-world scenarios, where factors such as atmospheric attenuation, temperature variations, vibrations, and other environmental effects can affect system performance. Some notable field demonstrations of MWP-based radar systems are summarized below.

In [137], an in-field demonstration of a photonic coherent MIMO distributed radar network was presented. The experiment verified the system's capability for long-range coherent signal distribution and demonstrated successful target detection in an outdoor environment. A fully photonics-based digital coherent radar system was demonstrated and tested in a real-world scenario in [138], confirming its ability to maintain high-resolution performance outside laboratory conditions. Additionally, a photonics-assisted radar system for target cluster detection was reported in [139]. This demonstration showcased the effectiveness of MWP technology in resolving closely spaced targets and confirmed the robustness of the photonic front-end under real environmental conditions.

These field tests show the practical feasibility of applying MWP technologies to radar systems, confirming key advantages such as wide-bandwidth operation, low phase noise, and high spatial resolution in non-ideal, real-world environments. Nevertheless, further field tests are needed to fully assess long-term reliability and performance under varying environmental and operational conditions.

4. Conclusions

MWP offers a promising solution to the limitations of traditional radar systems. Using the unique properties of light and optical components, MWP enables advances in radar performance, including increased bandwidth, reduced phase noise, improved EMI resilience, and enhanced spatial resolution.

This paper has explored key MWP techniques applicable to various radar functionalities. We have discussed how MWP can be used for signal transmission, LO signal generation, radar waveforming, optical beamforming networks, mixing, filtering, co-site interference suppression, RTFT, and ADC. These advances have the potential to revolutionize radar systems, enabling lower power consumption, operation in higher frequency bands, and greater precision.

While MWP presents exciting opportunities, challenges such as phase noise mitigation and system complexity remain. Ongoing research is focused on addressing these issues and further refining MWP technologies for radar applications. The potential for increased energy efficiency and enhanced resolution positions MWP as a key enabler for next-generation radar systems, promising to shape the future of radar technology.

Author Contributions: Conceptualization, L.P. and B.B.; Investigation, L.P.; Writing—original draft preparation, L.P.; Writing—review and editing, B.B., A.B. and A.L.; Visualization, L.P.; Supervision, B.B. All authors have read and agreed to the published version of the manuscript.

Funding: This work was supported by the Slovenian Research and Innovation Agency (ARIS) under grants J2-3048 and J2-50072 and research core funding No. P2-0246.

Conflicts of Interest: The authors declare no conflicts of interest.

References

1. Seeds, A.J.; Williams, K.J. Microwave Photonics. *J. Light. Technol.* **2006**, *24*, 4628–4641. [CrossRef]
2. Zhang, F.; Sun, G.; Zhou, Y.; Gao, B.; Pan, S. Towards High-Resolution Imaging With Photonics-Based Time Division Multiplexing MIMO Radar. *IEEE J. Sel. Top. Quantum Electron.* **2022**, *28*, 6000310.
3. De, S.; Raj, A.A.B. A survey on photonics technologies for radar applications. *J. Opt.* **2023**, *52*, 90–119. [CrossRef]
4. Blatnik, A.; Zmrzlak, L.; Batagelj, B. Radio Front-End for Frequency Agile Microwave Photonic Radars. *Electronics* **2024**, *13*. [CrossRef]
5. Nagatsuma, T.; Horiguchi, S.; Minamikata, Y.; Yoshimizu, Y.; Hisatake, S.; Kuwano, S.; Yoshimoto, N.; Terada, J.; Takahashi, H. Terahertz wireless communications based on photonics technologies. *Opt. Express* **2013**, *21*, 23736–23747. [CrossRef]
6. Pinna, S.; Melo, S.; Laghezza, F.; Scotti, F.; Lazzeri, E.; Scaffardi, M.; Ghelfi, P.; Bogoni, A. Photonics-Based Radar for Sub-mm Displacement Sensing. *IEEE J. Sel. Top. Quantum Electron.* **2017**, *23*, 168–175. [CrossRef]

7. Serafino, G.; Scotti, F.; Lembo, L.; Hussain, B.; Porzi, C.; Malacarne, A.; Maresca, S.; Onori, D.; Ghelfi, P.; Bogoni, A. Toward a New Generation of Radar Systems Based on Microwave Photonic Technologies. *J. Light. Technol.* **2019**, *37*, 643–650. [CrossRef]
8. Pan, S.; Zhang, Y. Microwave Photonic Radars. *J. Light. Technol.* **2020**, *38*, 5450–5484. [CrossRef]
9. Zhang, F.; Guo, Q.; Pan, S. Photonics-based real-time ultra-high-range-resolution radar with broadband signal generation and processing. *Sci. Rep.* **2017**, *7*, 13848. [CrossRef]
10. Harmuth, H.F. Frequency-Sharing and Spread-Spectrum Transmission with Large Relative Bandwidth. *IEEE Trans. Electromagn. Compat.* **1978**, *EMC-20*, 232–239. [CrossRef]
11. Waldschmidt, C.; Hasch, J.; Menzel, W. Automotive Radar — From First Efforts to Future Systems. *IEEE J. Microwaves* **2021**, *1*, 135–148. [CrossRef]
12. Yi, L.; Li, Y.; Nagatsuma, T. Photonic Radar for 3D Imaging: From Millimeter to Terahertz Waves. *IEEE J. Sel. Top. Quantum Electron.* **2023**, *29*, 8500714. [CrossRef]
13. Batra, A.; Barowski, J.; Damyanov, D.; Wiemeler, M.; Rolfes, I.; Schultze, T.; Balzer, J.C.; Göhringer, D.; Kaiser, T. Short-Range SAR Imaging From GHz to THz Waves. *IEEE J. Microwaves* **2021**, *1*, 574–585. [CrossRef]
14. Yi, L.; Nishida, Y.; Sagisaka, T.; Kaname, R.; Mizuno, R.; Fujita, M.; Nagatsuma, T. Towards Practical Terahertz Imaging System With Compact Continuous Wave Transceiver. *J. Light. Technol.* **2021**, *39*, 7850–7861. [CrossRef]
15. Tratnik, J.; Lemut, P.; Vidmar, M. Time-transfer and synchronization equipment for high-performance particle accelerators. *J. Microelectron. Electron. Components Mater.* **2012**, *42*, 115–122.
16. Koala, R.A.S.D.; Fujita, M.; Nagatsuma, T. Nanophotonics-inspired all-silicon waveguide platforms for terahertz integrated systems. *Nanophotonics* **2022**, *11*, 1741–1759. [CrossRef]
17. Strömbeck, F.; Yan, Y.; Zirath, H. A Beyond 100-Gbps Polymer Microwave Fiber Communication Link at D-Band. *IEEE Trans. Circuits Syst. I Regul. Pap.* **2023**, *70*, 3017–3028. [CrossRef]
18. Samarasekera, A.C.J.; Fernández, S.L.; Feger, R.; Hüttner, R.; Gruson, F.; Krainer, S.; Stelzer, A. Distributed Radar Network with Polymer Microwave Fiber (PMF) Based Synchronization. In Proceedings of the 2024 IEEE Topical Conference on Wireless Sensors and Sensor Networks (WiSNeT), San Antonio, TX, USA, 21–24 January 2024; pp. 1–4. [CrossRef]
19. Reynaert, P.; Dens, K.; D'heer, C.; Simic, D.; Vaes, J.; Philippe, B.; Ooms, S. Polymer Microwave Fiber: A New Communication Concept That Blends Wireless, Wireline and Optical Communication. In Proceedings of the 2019 26th IEEE International Conference on Electronics, Circuits and Systems (ICECS), Genoa, Italy, 27–29 November 2019; pp. 755–758. [CrossRef]
20. Dragonja, U.; Tratnik, J.; Batagelj, B. Use of copper-coated fiber as a tunable optical time-delay line in precise timing systems. *Opt. Quantum Electron.* **2013**, *45*, 1229–1235. [CrossRef]
21. Podbregar, L.; Dragonja, U.; Krč, J.; Czermann, M.; Blatnik, A.; Batagelj, B. Sensitivity of radio frequency phase stabilization system to temperature changes in a fiber optic link. In Proceedings of the 2024 7th International Balkan Conference on Communications and Networking (BalkanCom), Ljubljana, Slovenia , 3–6 June 2024; pp. 143–147. [CrossRef]
22. Wang, K.; Wei, W.; Wang, D.; Wang, P.; Xie, W.; Dong, Y. Distributed receiving system with local digitization and combination for SNR enhancement. *Opt. Express* **2023**, *31*, 1014–1024. [CrossRef]
23. Hu, C.; Luo, B.; Bai, W.; Pan, W.; Yan, L.; Zou, X. Stable Radio Frequency Transmission of Single Optical Source Over Fiber Based on Passive Phase Compensation. *IEEE Photonics J.* **2021**, *13*, 7200607. [CrossRef]
24. Tratnik, J.; Dragonja, U.; Batagelj, B. Multi-purpose constant-delay optical link. In Proceedings of the 2014 European Frequency and Time Forum (EFTF), Neuchatel, Switzerland, 23–26 June 2014; pp. 333–335. [CrossRef]
25. Ben-Amram, A.; Stern, Y.; London, Y.; Antman, Y.; Zadok, A. Stable closed-loop fiber-optic delay of arbitrary radio-frequency waveforms. *Opt. Express* **2015**, *23*, 28244–28257. [CrossRef]
26. Sun, S.; Petropulu, A.P.; Poor, H.V. MIMO Radar for Advanced Driver-Assistance Systems and Autonomous Driving: Advantages and Challenges. *IEEE Signal Process. Mag.* **2020**, *37*, 98–117. [CrossRef]
27. Kruse, S.; Gudyriev, S.; Schwabe, T.; Kneuper, P.; Kurz, H.G.; Scheytt, J.C. Silicon Photonic Radar Transmitter IC for mm-Wave Large Aperture MIMO Radar Using Optical Clock Distribution. *IEEE Microw. Wirel. Components Lett.* **2021**, *31*, 783–786. [CrossRef]
28. Kruse, S.; Gudyriev, S.; Kneuper, P.; Schwabe, T.; Meinecke, M.M.; Kurz, H.G.; Scheytt, J.C. Silicon Photonic Radar Receiver IC for mm-Wave Large Aperture MIMO Radar Using Optical Clock Distribution. *IEEE Microw. Wirel. Components Lett.* **2022**, *32*, 1447–1450. [CrossRef]
29. Ishio, H.; Minowa, J.; Nosu, K. Review and status of wavelength-division-multiplexing technology and its application. *J. Light. Technol.* **1984**, *2*, 448–463. [CrossRef]
30. Cvijetic, N.; Qian, D.; Yu, J.; Huang, Y.K.; Wang, T. Polarization-Multiplexed Optical Wireless Transmission With Coherent Detection. *J. Light. Technol.* **2010**, *28*, 1218–1227. [CrossRef]
31. Kikuchi, K. Performance analyses of polarization demultiplexing based on constant-modulus algorithm in digital coherent optical receivers. *Opt. Express* **2011**, *19*, 9868–9880. [CrossRef]
32. Chen, Z.Y.; Yan, L.S.; Pan, Y.; Jiang, L.; Yi, A.L.; Pan, W.; Luo, B. Use of polarization freedom beyond polarization-division multiplexing to support high-speed and spectral-efficient data transmission. *Light. Sci. Appl.* **2017**, *6*, e16207–e16207. [CrossRef]

33. Badraoui, N.; Berceli, T. Enhancing capacity of optical links using polarization multiplexing. *Opt. Quantum Electron.* **2019**, *51*, 310. [CrossRef]
34. Puttnam, B.J.; Rademacher, G.; Luís, R.S. Space-division multiplexing for optical fiber communications. *Optica* **2021**, *8*, 1186–1203. [CrossRef]
35. Majumder, S.P.; Hossain, A. Performance analysis of an SDM-WDM multicore optical fiber link. *Opt. Contin.* **2023**, *2*, 600–615. [CrossRef]
36. Berdagué, S.; Facq, P. Mode division multiplexing in optical fibers. *Appl. Opt.* **1982**, *21*, 1950–1955. [CrossRef] [PubMed]
37. Hanzawa, N.; Saitoh, K.; Sakamoto, T.; Matsui, T.; Tomita, S.; Koshiba, M. Demonstration of mode-division multiplexing transmission over 10 km two-mode fiber with mode coupler. In Proceedings of the 2011 Optical Fiber Communication Conference and Exposition and the National Fiber Optic Engineers Conference, Los Angeles, CA, USA, 6–10 March 2011; pp. 1–3.
38. Ryf, R.; Randel, S.; Gnauck, A.H.; Bolle, C.; Sierra, A.; Mumtaz, S.; Esmaeelpour, M.; Burrows, E.C.; Essiambre, R.J.; Winzer, P.J.; et al. Mode-Division Multiplexing Over 96 km of Few-Mode Fiber Using Coherent 6 × 6 MIMO Processing. *J. Light. Technol.* **2012**, *30*, 521–531. [CrossRef]
39. Sillard, P.; Bigot-Astruc, M.; Molin, D. Few-Mode Fibers for Mode-Division-Multiplexed Systems. *J. Light. Technol.* **2014**, *32*, 2824–2829. [CrossRef]
40. Rademacher, G.; Puttnam, B.J.; Luís, R.S.; Sakaguchi, J.; Klaus, W.; Eriksson, T.A.; Awaji, Y.; Hayashi, T.; Nagashima, T.; Nakanishi, T.; et al. 10.66 Peta-Bit/s Transmission over a 38-Core-Three-Mode Fiber. In Proceedings of the 2020 Optical Fiber Communications Conference and Exhibition (OFC), San Diego, CA, USA, 8–12 March 2020; pp. 1–3.
41. Maher, R.; Xu, T.; Galdino, L.; Sato, M.; Alvarado, A.; Shi, K.; Savory, S.J.; Thomsen, B.C.; Killey, R.I.; Bayvel, P. Spectrally Shaped DP-16QAM Super-Channel Transmission with Multi-Channel Digital Back-Propagation. *Sci. Rep.* **2015**, *5*, 8214. [CrossRef]
42. Luo, L.W.; Ophir, N.; Chen, C.P.; Gabrielli, L.H.; Poitras, C.B.; Bergmen, K.; Lipson, M. WDM-compatible mode-division multiplexing on a silicon chip. *Nat. Commun.* **2014**, *5*, 3069. [CrossRef]
43. Yao, J. Microwave Photonics: Photonic Generation of Microwave and Millimeter-wave Signals. *Int. J. Microw. Opt. Technol.* **2010**, *5*.
44. Wei, J.; Zhang, S.; Kim, J.; Pan, S. Compact Phase Detector for Optical-Microwave Synchronization Using Polarization Modulation. *J. Light. Technol.* **2018**, *36*, 4267–4272. [CrossRef]
45. Ponnampalam, L.; Fice, M.J.; Pozzi, F.; Renaud, C.C.; Rogers, D.C.; Lealman, I.F.; Moodie, D.G.; Cannard, P.J.; Lynch, C.; Johnston, L.; et al. Monolithically Integrated Photonic Heterodyne System. *J. Light. Technol.* **2011**, *29*, 2229–2234. [CrossRef]
46. Steed, R.J.; Ponnampalam, L.; Fice, M.J.; Renaud, C.C.; Rogers, D.C.; Moodie, D.G.; Maxwell, G.D.; Lealman, I.F.; Robertson, M.J.; Pavlovic, L.; et al. Hybrid Integrated Optical Phase-Lock Loops for Photonic Terahertz Sources. *IEEE J. Sel. Top. Quantum Electron.* **2011**, *17*, 210–217. [CrossRef]
47. Zhang, F.; Pan, S. Microwave photonic signal generation for radar applications. In Proceedings of the 2016 IEEE International Workshop on Electromagnetics: Applications and Student Innovation Competition (iWEM), Nanjing, China, 16–18 May 2016; pp. 1–3. [CrossRef]
48. Zhou, W.; Blasche, G. Injection-locked dual opto-electronic oscillator with ultra-low phase noise and ultra-low spurious level. *IEEE Trans. Microw. Theory Tech.* **2005**, *53*, 929–933. [CrossRef]
49. Hao, T.; Liu, Y.; Tang, J.; Cen, Q.; Li, W.; Zhu, N.; Dai, Y.; Capmany, J.; Yao, J.; Li, M. Recent advances in optoelectronic oscillators. *Adv. Photonics* **2020**, *2*, 044001. [CrossRef]
50. Bogataj, L.; Vidmar, M.; Batagelj, B. A Feedback Control Loop for Frequency Stabilization in an Opto-Electronic Oscillator. *J. Light. Technol.* **2014**, *32*, 3690–3694. [CrossRef]
51. Batagelj, B.; Bogataj, L.; Vidmar, M. Key properties and design issues for an opto-electronic oscillator. In Proceedings of the 2015 17th International Conference on Transparent Optical Networks (ICTON), Budapest, Hungary , 5–9 July 2015; pp. 1–4. [CrossRef]
52. Bogataj, L.; Vidmar, M.; Batagelj, B. Improving the Side-mode Suppression Ratio and Reducing the Frequency Drift in an Opto-Electronic Oscillator With a Feedback Control Loop and Additional Phase Modulation. *J. Light. Technol.* **2016**, *34*, 885–890. [CrossRef]
53. Bogataj, L.; Vidmar, M.; Batagelj, B. Opto-Electronic Oscillator With Quality Multiplier. *IEEE Trans. Microw. Theory Tech.* **2016**, *64*, 663–668. [CrossRef]
54. Li, M.; Hao, T.; Yang, Y.; Shi, N.; Li, W. Recent advances in optoelectronic oscillators and quantum microwave photonics. In Proceedings of the 2021 International Topical Meeting on Microwave Photonics (MWP), Pisa, Italy, 15–17 November 2021; pp. 1–3. [CrossRef]
55. Ilgaz, M.A.; Batagelj, B. Opto-Electronic Oscillators for Micro- and Millimeter Wave Signal Generation. *Electronics* **2021**, *10*, 857. [CrossRef]
56. Fu, Z.; Ma, Y.; Zeng, Z.; Zhang, L.; Tian, H.; Lyu, W.; Zhang, Y.; Zhang, Z.; Zhang, S.; Li, H.; et al. Microwave local oscillator signal generation with low near-carrier phase noise based on cascaded optoelectronic oscillators. *Opt. Laser Technol.* **2025**, *181*, 111628. [CrossRef]

57. Liu, Q.; Peng, J.; Yan, J. Optoelectronic Oscillators: Progress from Classical Designs to Integrated Systems. *Photonics* **2025**, *12*, 120. [CrossRef]
58. Qi, G.; Yao, J.; Seregelyi, J.; Paquet, S.; Belisle, C. Generation and distribution of a wide-band continuously tunable millimeter-wave signal with an optical external modulation technique. *IEEE Trans. Microw. Theory Tech.* **2005**, *53*, 3090–3097. [CrossRef]
59. Lavrič, A.; Batagelj, B. Phase Noise Measurement for Improving the Performance of High-Order Modulation Scheme Wireless Links. In Proceedings of the 2024 7th International Balkan Conference on Communications and Networking (BalkanCom), Ljubljana, Slovenia, 3–6 June 2024; pp. 148–152. [CrossRef]
60. Stöhr, A.; Babiel, S.; Cannard, P.J.; Charbonnier, B.; van Dijk, F.; Fedderwitz, S.; Moodie, D.; Pavlovic, L.; Ponnampalam, L.; Renaud, C.C.; et al. Millimeter-Wave Photonic Components for Broadband Wireless Systems. *IEEE Trans. Microw. Theory Tech.* **2010**, *58*, 3071–3082. [CrossRef]
61. Luo, H.; Yu, J.; Wang, J.; Ma, C.; Han, X.; Su, X.; Gao, Y.; Jia, S. Microwave Photonic Frequency Multiplier with Low Phase Noise Based on an Optoelectronic Oscillator. *Photonics* **2024**, *11*, 588. [CrossRef]
62. Peng, M.Y.; Kalaydzhyan, A.; Kärtner, F.X. Balanced optical-microwave phase detector for sub-femtosecond optical-RF synchronization. *Opt. Express* **2014**, *22*, 27102–27111. [CrossRef] [PubMed]
63. Hasanuzzaman, G.K.M.; Iezekiel, S.; Kanno, A. W-Band Optoelectronic Oscillator. *IEEE Photonics Technol. Lett.* **2020**, *32*, 771–774. [CrossRef]
64. Luchinin, A.; Malygin, I. Optoelectronic Microwave Oscillator. Advantages and Disadvantages. Noise Sources in an OEO. In Proceedings of the 2024 IEEE Ural-Siberian Conference on Biomedical Engineering, Radioelectronics and Information Technology (USBEREIT), Yekaterinburg, Russia, 13–15 May 2024; pp. 333–337. [CrossRef]
65. Kaba, M.; Li, H.W.; Daryoush, A.; Vilcot, J.P.; Decoster, D.; Chazelas, J.; Bouwmans, G.; Quiquempois, Y.; Deborgies, F. Improving thermal stability of opto-electronic oscillators. *IEEE Microw. Mag.* **2006**, *7*, 38–47. [CrossRef]
66. Zhu, X.; Wei, W.; Xie, W.; Dong, Y. Long-Term Stable Optoelectronic Oscillator Based on Phase-Locked Loop. In Proceedings of the 2023 IEEE 7th International Symposium on Electromagnetic Compatibility (ISEMC), Hangzhou, China, 20–23 October 2023; pp. 1–3. [CrossRef]
67. Eliyahu, D.; Sariri, K.; Kamran, A.; Tokhmakhian, M. Improving short and long term frequency stability of the opto-electronic oscillator. In Proceedings of the 2002 IEEE International Frequency Control Symposium and PDA Exhibition (Cat. No.02CH37234), New Orleans, LA, USA, 31 May 2002; pp. 580–583. [CrossRef]
68. Wang, C.; Yao, J. Photonic Generation of Chirped Microwave Pulses Using Superimposed Chirped Fiber Bragg Gratings. *IEEE Photonics Technol. Lett.* **2008**, *20*, 882–884. [CrossRef]
69. Li, M.; Shao, L.Y.; Albert, J.; Yao, J. Tilted Fiber Bragg Grating for Chirped Microwave Waveform Generation. *IEEE Photonics Technol. Lett.* **2011**, *23*, 314–316. [CrossRef]
70. Rashidinejad , A.; Weiner, A.M. Photonic Radio-Frequency Arbitrary Waveform Generation With Maximal Time-Bandwidth Product Capability. *J. Light. Technol.* **2014**, *32*, 3383–3393. [CrossRef]
71. Zhou, P.; Zhang, F.; Guo, Q.; Pan, S. Linearly chirped microwave waveform generation with large time-bandwidth product by optically injected semiconductor laser. *Opt. Express* **2016**, *24*, 18460–18467. [CrossRef]
72. Zhou, P.; Zhang, F.; Guo, Q.; Li, S.; Pan, S. Reconfigurable Radar Waveform Generation Based on an Optically Injected Semiconductor Laser. *IEEE J. Sel. Top. Quantum Electron.* **2017**, *23*, 1801109. [CrossRef]
73. Nakarmi, B.; Chen, H.; Ukaegbu, I.A.; Gaudel, B.; Nakarmi, U.; Muthaiyah, S.; Ng, A.; Wang, X.; Pan, S. Linear Frequency Modulated Photonics RADAR using Injection Locking in Semiconductor Laser. In Proceedings of the 2021 International Conference on Electronics, Communications and Information Technology (ICECIT), Khulna, Bangladesh, 14–16 September 2021; pp. 1–4. [CrossRef]
74. Kumar, C.; Raghuwanshi, S.K.; Yadav, R.K. A review on photonic-assisted dual-chirp microwave waveform generation methods. *Microw. Opt. Technol. Lett.* **2024**, *66*, e34266. [CrossRef]
75. Yao, Y.; Zhang, F.; Zhang, Y.; Ye, X.; Zhu, D.; Pan, S. Demonstration of Ultra-high-resolution Photonics-based Kaband Inverse Synthetic Aperture Radar Imaging. In Proceedings of the 2018 Optical Fiber Communications Conference and Exposition (OFC), San Diego, CA, USA, 11–15 March 2018; pp. 1–3.
76. Guo, Q.; Zhang, F.; Zhou, P.; Pan, S. Dual-Band LFM Signal Generation by Optical Frequency Quadrupling and Polarization Multiplexing. *IEEE Photonics Technol. Lett.* **2017**, *29*, 1320–1323. [CrossRef]
77. Zhang, F.; Gao, B.; Pan, S. Arbitrary waveform generation by a 4-bit photonic digital-to-analog converter with 3.49 effective number of bits. In Proceedings of the 2016 IEEE International Topical Meeting on Microwave Photonics (MWP), Long Beach, CA, USA, 31 October–3 November 2016; pp. 137–140. [CrossRef]
78. Peng, S.; Li, S.; Xue, X.; Xiao, X.; Wu, D.; Zheng, X.; Zhou, B. High-resolution W-band ISAR imaging system utilizing a logic-operation-based photonic digital-to-analog converter. *Opt. Express* **2018**, *26*, 1978–1987. [CrossRef]

79. Li, J.; Xue, X.; Zha, Y.; Wu, D.; Li, S.; Zheng, X. A Segmented Photonic Digital-to-analog Converter with a High Effective Number of Bits. In Proceedings of the 2019 International Topical Meeting on Microwave Photonics (MWP), Ottawa, ON, Canada , 7–10 October 2019; pp. 1–3. [CrossRef]
80. García, S.; Ureña, M.; Gasulla, I. Photonic-Assisted Analog-to-Digital Conversion Based on a Dispersion-Diversity Multicore Fiber. *IEEE J. Sel. Top. Quantum Electron.* **2023**, *29*, 4300108. [CrossRef]
81. Liu, Z.; Slavík, R. Optical Injection Locking: From Principle to Applications. *J. Light. Technol.* **2020**, *38*, 43–59. [CrossRef]
82. Won, Y.S.; Kim, C.H.; Lee, S.G. Range Resolution Improvement of a 24 GHz ISM Band Pulse Radar—A Feasibility Study. *IEEE Sens. J.* **2015**, *15*, 7142–7149. [CrossRef]
83. Vokić, N.; Milovančev, D. Photomixing Techniques for mm-Wave Carrier Generation. In Proceedings of the 2024 47th MIPRO ICT and Electronics Convention (MIPRO), Opatija, Croatia, 20–24 May 2024; pp. 1671–1676. [CrossRef]
84. Minasian, R.A.; Chan, E.H.W.; Yi, X. Microwave photonic signal processing. *Opt. Express* **2013**, *21*, 22918–22936. [CrossRef] [PubMed]
85. Zhou, Y.; Kong, J.; Zhang, F.; Pan, S. Microwave photonic I/Q mixer for wideband frequency downconversion with serial electro-optical modulations. *Opt. Lett.* **2024**, *49*, 65–68. [CrossRef]
86. Tang, Z.; Li, Y.; Yao, J.; Pan, S. Photonics-Based Microwave Frequency Mixing: Methodology and Applications. *Laser Photonics Rev.* **2020**, *14*, 1800350. [CrossRef]
87. Kittlaus, E.A.; Eliyahu, D.; Ganji, S.; Williams, S.; Matsko, A.B.; Cooper, K.B.; Forouhar, S. A low-noise photonic heterodyne synthesizer and its application to millimeter-wave radar. *Nat. Commun.* **2021**, *12*, 4397. [CrossRef]
88. Yao, J. Microwave Photonics. *J. Light. Technol.* **2009**, *27*, 314–335. [CrossRef]
89. Zhu, D.; Pan, S. Photonics-Based Microwave Image-Reject Mixer. *Photonics* **2018**, *5*, 6. [CrossRef]
90. Capmany, J.; Mora, J.; Gasulla, I.; Sancho, J.; Lloret, J.; Sales, S. Microwave Photonic Signal Processing. *J. Light. Technol.* **2013**, *31*, 571–586. [CrossRef]
91. Han, X.; Huo, B.; Shao, Y.; Wang, C.; Zhao, M. RF Self-Interference Cancellation Using Phase Modulation and Optical Sideband Filtering. *IEEE Photonics Technol. Lett.* **2017**, *29*, 917–920. [CrossRef]
92. Chen, Y.; Pan, S. Photonics-Assisted Radio-Frequency Self-Interference Cancellation and Fiber Transmission Using a DP-QPSK modulator. In Proceedings of the 2018 International Topical Meeting on Microwave Photonics (MWP), Toulouse, France, 22–25 October 2018; pp. 1–4. [CrossRef]
93. Zhou, Y.; Wang, L.; Liu, Y.; Yu, Y.; Zhang, X. Microwave Photonic Filters and Applications. *Photonics* **2023**, *10*, 1110. [CrossRef]
94. Tao, Z.; Tao, Y.; Jin, M.; Qin, J.; Chen, R.; Shen, B.; Wu, Y.; Shu, H.; Yu, S.; Wang, X. Highly reconfigurable silicon integrated microwave photonic filter towards next-generation wireless communication. *Photon. Res.* **2023**, *11*, 682–694. [CrossRef]
95. Porzi, C.; Serafino, G.; Velha, P.; Ghelfi, P.; Bogoni, A. Integrated SOI High-Order Phase-Shifted Bragg Grating for Microwave Photonics Signal Processing. *J. Light. Technol.* **2017**, *35*, 4479–4487. [CrossRef]
96. Tu, Z.; Wen, A.; Li, X.; Zhang, H. A Photonic Pre-Distortion Technique for RF Self-Interference Cancellation. *IEEE Photonics Technol. Lett.* **2018**, *30*, 1297–1300. [CrossRef]
97. Shi, N.; Song, Q.; Tang, J.; Li, W.; Zhu, N.; Li, M. A switchable self-interference cancellation system for dual-band IBFD system using a monolithic integrated DML array. *Opt. Commun.* **2019**, *447*, 55–60. [CrossRef]
98. Wang, H.; Wen, A.; Men, Y.; Wang, Y. Photonics-Based Radar Jamming Signal Generation and Self-Interference Cancellation. *J. Light. Technol.* **2023**, *41*, 5922–5929. [CrossRef]
99. Bonjour, R.; Singleton, M.; Leuchtmann, P.; Leuthold, J. Comparison of steering angle and bandwidth for various phased array antenna concepts. *Opt. Commun.* **2016**, *373*, 35–43. SI: Microwave photonics. [CrossRef]
100. Abdalrazak, M.Q.; Majeed, A.H.; Abd-Alhameed, R.A. A Critical Examination of the Beam-Squinting Effect in Broadband Mobile Communication: Review Paper. *Electronics* **2023**, *12*, 400. [CrossRef]
101. Cheng, Q.; Zheng, S.; Zhang, Q.; Ji, J.; Yu, H.; Zhang, X. An integrated optical beamforming network for two-dimensional phased array radar. *Opt. Commun.* **2021**, *489*, 126809. [CrossRef]
102. Duan, F.; Guo, Y.; Gu, Z.; Yin, Y.; Wu, Y.; Chen, T. Optical Beamforming Networks for Millimeter-Wave Wireless Communications. *Appl. Sci.* **2023**, *13*, 8346. [CrossRef]
103. Ritoša, P.; Batagelj, B.; Vidmar, M. Optically steerable antenna array for radio over fibre transmission. *Electron. Lett.* **2005**, *41*, 917–918. [CrossRef]
104. Ritosa, P.; Vidmar, M. Remote optically controllable antenna array using a single optical fiber for RoF reception. *Microw. Opt. Technol. Lett.* **2007**, *49*, 1349–1353. [CrossRef]
105. Shin, J.D.; Lee, B.S.; Kim, B.G. Optical true time-delay feeder for X-band phased array antennas composed of 2/spl times/2 optical MEMS switches and fiber delay lines. *IEEE Photonics Technol. Lett.* **2004**, *16*, 1364–1366. [CrossRef]
106. Matthews, P.; Liu, P.L.; Medberry, J.; Franekl, M.; Esman, R. Demonstration of a wide-band fiber-optic nulling system for array antennas. *IEEE Trans. Microw. Theory Tech.* **1999**, *47*, 1327–1331. [CrossRef]

107. Ye, X.; Zhang, F.; Pan, S. Compact optical true time delay beamformer for a 2D phased array antenna using tunable dispersive elements. *Opt. Lett.* **2016**, *41*, 3956–3959. [CrossRef]
108. Luan, Y.; Yang, T.; Ren, J.; Li, R.; Zhang, Z. Broadband Beam-Scanning Phased Array Based on Microwave Photonics. *Electronics* **2024**, *13*, 1278. [CrossRef]
109. Yi, X.; Li, L.; Huang, T.X.H.; Minasian, R.A. Programmable multiple true-time-delay elements based on a Fourier-domain optical processor. *Opt. Lett.* **2012**, *37*, 608–610. [CrossRef]
110. Cardenas, J.; Foster, M.A.; Sherwood-Droz, N.; Poitras, C.B.; Lira, H.L.R.; Zhang, B.; Gaeta, A.L.; Khurgin, J.B.; Morton, P.; Lipson, M. Wide-bandwidth continuously tunable optical delay line using silicon microring resonators. *Opt. Express* **2010**, *18*, 26525–26534. [CrossRef]
111. Hong, S.; Zhang, L.; Wang, Y.; Zhang, M.; Xie, Y.; Dai, D. Ultralow-loss compact silicon photonic waveguide spirals and delay lines. *Photon. Res.* **2022**, *10*, 1–7. [CrossRef]
112. Hong, S.; Zhang, L.; Wu, J.; Peng, Y.; Lyu, L.; Hu, Y.; Xie, Y.; Dai, D. Multimode-enabled silicon photonic delay lines: Break the delay-density limit. *Light Sci. Appl.* **2015**, *14*, 145. [CrossRef] [PubMed]
113. Jannson, T. Real-time Fourier transformation in dispersive optical fibers. *Opt. Lett.* **1983**, *8*, 232–234. [CrossRef] [PubMed]
114. van Howe, J.; Xu, C. Ultrafast Optical Signal Processing Based Upon Space-Time Dualities. *J. Light. Technol.* **2006**, *24*, 2649. [CrossRef]
115. Malacarne, A.; Park, Y.; Li, M.; LaRochelle, S.; Azaña, J. Real-time Fourier transformation of lightwave spectra and application in optical reflectometry. *Opt. Express* **2015**, *23*, 32516–32527. [CrossRef]
116. Zheng, Y.; Li, J.; Dai, Y.; Yin, F.; Xu, K. Real-time Fourier transformation based on the bandwidth magnification of RF signals. *Opt. Lett.* **2018**, *43*, 194–197. [CrossRef]
117. Li, J.; Fu, S.; Xie, X.; Xiang, M.; Dai, Y.; Yin, F.; Qin, Y. Low-Latency Short-Time Fourier Transform of Microwave Photonics Processing. *J. Light. Technol.* **2023**, *41*, 6149–6156. [CrossRef]
118. Wang, C. Dispersive Fourier Transformation for Versatile Microwave Photonics Applications. *Photonics* **2014**, *1*, 586–612. [CrossRef]
119. Akca, B.I. Design of a compact and ultrahigh-resolution Fourier-transform spectrometer. *Opt. Express* **2017**, *25*, 1487–1494. [CrossRef]
120. Souza, M.C.M.M.; Grieco, A.; Frateschi, N.C.; Fainman, Y. Fourier transform spectrometer on silicon with thermo-optic non-linearity and dispersion correction. *Nat. Commun.* **2018**, *9*, 665. [CrossRef]
121. Li, A.; Yao, C.; Xia, J.; Wang, H.; Cheng, Q.; Penty, R.; Fainman, Y.; Pan, S. Advances in cost-effective integrated spectrometers. *Light Sci .Appl.* **2022**, *11*, 174. [CrossRef] [PubMed]
122. Salem, R.; Foster, M.A.; Gaeta, A.L. Application of space–time duality to ultrahigh-speed optical signal processing. *Adv. Opt. Photon.* **2013**, *5*, 274–317. [CrossRef]
123. Yang, D.; Kumar, S. Realization of optical OFDM using time lenses and its comparison with optical OFDM using FFT. *Opt. Express* **2009**, *17*, 17214–17226. [CrossRef] [PubMed]
124. Li, M.; Wang, C.; Li, W.; Yao, J. An Unbalanced Temporal Pulse-Shaping System for Chirped Microwave Waveform Generation. *IEEE Trans. Microw. Theory Tech.* **2010**, *58*, 2968–2975. [CrossRef]
125. de Chatellus, H.G.; Cortés, L.R.; Azaña, J.A. Optical real-time Fourier transformation with kilohertz resolutions. *Optica* **2016**, *3*, 1–8. [CrossRef]
126. Valley, G.C. Photonic analog-to-digital converters. *Opt. Express* **2007**, *15*, 1955–1982. [CrossRef] [PubMed]
127. Masnad, M.M.; Safaee, S.M.R.; Najeeb, N.; Mojaver, K.R.; Fouda, M.; Peinke, E.; Liboiron-Ladouceur, O. Scalable Photonic Digital-to-Analog Converters. *Photonics* **2024**, *11*, 112. [CrossRef]
128. Han, Y.; Guo, Y.; Gao, B.; Ma, C.; Zhang, R.; Zhang, H. Generation, optimization, and application of ultrashort femtosecond pulse in mode-locked fiber lasers. *Prog. Quantum Electron.* **2020**, *71*, 100264. [CrossRef]
129. Laghezza, F.; Scotti, F.; Ghelfi, P.; Bogoni, A.; Pinna, S. Jitter-limited photonic analog-to-digital converter with 7 effective bits for wideband radar applications. In Proceedings of the 2013 IEEE Radar Conference (RadarCon13), Ottawa, ON, Canada, 29 April–3 May 2013; pp. 1–5. [CrossRef]
130. Ghelfi, P.; Laghezza, F.; Scotti, F.; Serafino, G.; Capria, A.; Pinna, S.; Onori, D.; Porzi, C.; Scaffardi, M.; Malacarne, A.; et al. A fully photonics-based coherent radar system. *Nature* **2014**, *507*, 341–345. [CrossRef]
131. Fard, A.M.; Gupta, S.; Jalali, B. Photonic time-stretch digitizer and its extension to real-time spectroscopy and imaging. *Laser Photonics Rev.* **2013**, *7*, 207–263. [CrossRef]
132. Zou, W.; Qian, N.; Qin, R.; Wang, J.; Xu, S.; Zou, X.; Zhou, D. Integrated Photonic Analog-to-Digital Converter and Its Applications. In Proceedings of the 2024 International Topical Meeting on Microwave Photonics (MWP), Pisa, Italy, 17–20 September 2024; pp. 1–4. [CrossRef]

133. Fang, D.; Zazzi, A.; Müller, J.; Drayss, D.; Füllner, C.; Marin-Palomo, P.; Tabatabaei Mashayekh, A.; Dipta Das, A.; Weizel, M.; Gudyriev, S.; et al. Optical Arbitrary Waveform Measurement (OAWM) Using Silicon Photonic Slicing Filters. *J. Light. Technol.* **2022**, *40*, 1705–1717. [CrossRef]
134. Fang, D.; Drayss, D.; Chen, Y.; Lauermann, M.; Peng, H.; Lihachev, G.; Quint, A.; Valenziano, L.; Randel, S.; Zwick, T.; et al. Spectrally Sliced Optical Arbitrary Waveform Measurement (OAWM) Using a Photonic Multi-Chip Receiver Assembly. In Proceedings of the 2024 Optical Fiber Communications Conference and Exhibition (OFC), San Diego, CA, USA, 24–28 March 2024; pp. 1–3.
135. Li, Z.; Wang, X.; Zhang, Y.; Shang, C.; Lyu, W.; Lyu, Y.; Zeng, C.; Zhang, Z.; Zhang, S.; Li, H.; et al. Photonic sampling analog-to-digital conversion based on time and wavelength interleaved ultra-short optical pulse train generated by using monolithic integrated LNOI intensity and phase modulator. *Opt. Express* **2022**, *30*, 29611–29620. [CrossRef] [PubMed]
136. Wang, X.; Zhou, G.; Guo, Y.; Lu, L.; Nisar, M.S.; Chen, J.; Zhou, L. Integrated High-Repetition-Rate Optical Sampling Chip Exploiting Wavelength and Mode Multiplexing. *J. Light. Technol.* **2021**, *39*, 5548–5557. [CrossRef]
137. Lembo, L.; Maresca, S.; Serafino, G.; Scotti, F.; Amato, F.; Ghelfi, P.; Bogoni, A. In-Field Demonstration of a Photonic Coherent MIMO Distributed Radar Network. In Proceedings of the 2019 IEEE Radar Conference (RadarConf), Boston, MA, USA, 22–26 April 2019; pp. 1–6. [CrossRef]
138. Scotti, F.; Laghezza, F.; Serafino, G.; Pinna, S.; Onori, D.; Ghelfi, P.; Bogoni, A. In-Field Experiments of the First Photonics-Based Software-Defined Coherent Radar. *J. Light. Technol.* **2014**, *32*, 3365–3372. [CrossRef]
139. Zhou, Y.; Zhang, F.; Kong, J.; Wang, Y.; Li, J.; Chen, K.; Sun, G.; He, Y.; Pan, S. Field Trial of Cluster Target Detection by Broadband Microwave Photonic MIMO Radar. *IEEE J. Microwaves* **2025**, *5*, 631–639. [CrossRef]

MDPI AG
Grosspeteranlage 5
4052 Basel
Switzerland
Tel.: +41 61 683 77 34

Photonics Editorial Office
E-mail: photonics@mdpi.com
www.mdpi.com/journal/photonics

www.ingramcontent.com/pod-product-compliance
Lightning Source LLC
LaVergne TN
LVHW070445061126
851858LV00011B/164